US ARMY INSIDER MISSIONS 3
Nordic ETs, Space Arks & Saturn

US ARMY INSIDER MISSIONS 3
Nordic ETs, Space Arks & Saturn

MICHAEL E. SALLA, PH.D.
BOOK THREE OF THE US ARMY INSIDER MISSIONS SERIES

KEY WEST, FLORIDA, USA

US ARMY INSIDER MISSIONS 3
NORDIC ETS, SPACE ARKS & SATURN

Copyright © 2024 by Michael E. Salla, Ph.D.

All Rights Reserved. No part of this book may be reproduced or translated into any language or utilized in any form or by any means, electronic or mechanical, including photocopying, recording or by any information storage and retrieval system, without written permission by the author.

Exopolitics Today
PO Box 2672
Key West, FL 33040 USA

Printed in the United States of America

Cover Design: Adam S. Doyle

ISBN 979-8-9891404-3-5

Dedication

To those unnamed US military officers and "White Hats" who have offered protection and counsel to JP over the years of his updates and military service. You have performed a profound service to humanity's awakening to our Inner Earth and galactic community.

Table of Contents

Dedication v

Table of Figures ix

Preface 1

Chapter 1 5

Returning the Activation Jewel to the Atlantic Space Ark 5

   Interview Transcript with Commentary 5

Chapter 2 37

Mission to Nordic Inner Earth Civilization for Life Extension Technology 37

Chapter 3 61

Inside a Flying Saucer and Nordics take control of Space Arks .. 61

   Nordics Take Control of the Space Arks 72

Chapter 4 81

6th Mission to Atlantic Space Ark 81

Chapter 5 111

Strange Medical Experiments & Atlantic Ark Update 111

   Strange Medical Experiments 111

   More on Nordics Taking Over the Space Arks 128

Chapter 6 145

Nordic Assimilation Facility 145

Chapter 7 177

The Ancient Underground Castle with Gold Plates 177

Chapter 8 203

Ant People Move to New Realm after Ant King Transitions & Sleeping Giant Awakens 203

Chapter 9 233

Nordics Training International Military Pilots to fly Flying Saucers 233

Chapter 10 261

Testing for a Covert Space Mission to Saturn 261

Chapter 11 281

Roundtable on Space Arks, Sleeping Giants, ET Assimilation & Mysteries of Saturn 281

Chapter 12 327

Honorable Discharge from US Army and Future Covert Missions 327

Acknowledgments 337

About the Author 339

Also by DR. MICHAEL SALLA 341

Index 343

Endnotes 347

Table of Figures

Figure 1. NOAA map tracking Hurricane Lee's Path. NOAA. 8
Figure 2. World ocean currents. Wikipedia: Public Domain 10
Figure 3. Underwater complex of three domes that Jean Charles Moyen witnessed and drew in 1979. 20
Figure 4. Open Source Intelligence. Statista 24
Figure 5. Dragon's Triangle, aka Devil's Triangle. Source: Google Maps 28
Figure 6. Mesa Verde - Pueblo land. By Gleb Tarassenko 49
Figure 7 JP photo of a flying saucer showing two figures behind a rectangular porthole 63
Figure 8. Illustration of Orthon, who met with George Adamski in 1952. 66
Figure 9. Photo of Flying Triangle 84
Figure 10. Narkina 5 Prison Complex appeared in Disney Channel's Andor - Star Wars TV series 101
Figure 11. Senator Frank Church holds up the "heart attack gun." 104
Figure 12 JP took a selfie while hooked up to electrodes during the medical experiment 118
Figure 13. Stephen Hawking speaking via Hologram in Hong Kong 120
Figure 14. List of Parties to United Nations Convention on the Law of the Sea. Source: Wikimedia Commons 135
Figure 15 List of Parties to Underwater Cultural Heritage Convention. Source: Wikimedia Commons 139
Figure 16. The Phobos monolith - Mars Global Surveyor/NASA. 143
Figure 17. Photos of a craft similar to the one in which two Nordics from the Moon arrived 152
Figure 18. Departing spacecraft carrying Nordic JP in USAF uniform 159
Figure 19. JP illustrated the scene he witnessed when arriving at the ice cavern 183

Figure 20. Sagrada Familia, Barcelona. Source: Adobe Stock 186
Figure 21. JP's illustration of the scene when leaving the giant ice cavern.. 198
Figure 22. Recipe for Chia Pomegranate Drink. Permission: Dine D'ávilla... 212
Figure 23 Celestial phenomenon over Nuremberg, Germany, on April 14, 1561. Illustrated by Hanns Glaser 224
Figure 24. Recipe for Immunity given to JP by Ant People. Permission: Dine D'ávilla 226
Figure 25. Alabama Cave Trail shows extensive caves in Northern Alabama 235
Figure 26. A day-time photo from 2017 where JP previously saw a flying triangle & helicopter together.. 255
Figure 27. Elena Dannan communication on JP and a future Saturn mission 320
Figure 28. Spacecraft identified around the A, B, C & D Saturnian rings. Ringmakers of Saturn, p. 50. 324
Figure 29. Certificate of Retirement... 330
Figure 30. Navy and Marine Corps Achievement Medal............ 331
Figure 31. DD Form 214 332

Preface

In *US Army Insider Missions 3*, JP's extraordinary covert missions as an active US Army soldier continue with him returning to a submerged space ark in the Atlantic Ocean with human-looking 'Nordic' extraterrestrials who are carrying a crystal jewel embedded with the ark's organic consciousness. Unknown parties have been using the jewel to activate ancient technologies and other space arks found worldwide. JP later meets a Nordic called 'Jafis' who gives him a strange blue drink that alleviates some of the accumulating health issues he has been experiencing due to his off-planet covert missions and uniquely hazardous army service. Jafis informs JP that the Nordics have taken over the Atlantic and other space arks to further activate and protect them as humanity's geo-political crises worsen.

In his sixth mission to the Atlantic Space Ark, JP explains how the Nordics have succeeded in activating the space arks due to their accumulated knowledge and greater experience with this type of technology. The Nordics are acting in cooperation with an international military coalition that is observing and learning about the space ark's advanced technologies.

Inside a major military base in Florida, JP and his team are shown a deep entry shaft that leads directly to an Inner Earth civilization. After descending on two separate elevators operating at harrowing speeds and passing different security stations with stringent protocols, he meets with Nordic-looking humans who escort him, a medical officer, and the rest of his team to an ancient library. The medical officer is given life extension information that the team takes back to the surface. Presumably, such information will be used for developing life extension technologies that will eventually be introduced to the rest of humanity.

On a mission with far-reaching implications, JP is part of a team of four soldiers who meet and escort four Nordic extraterrestrials to a remarkable facility at a major military base in

Florida. The Nordics enter the facility and come out with very different physical appearances, clothing, and documentation, which makes them indistinguishable from ordinary humans. The infiltration of human society by Nordic and other visiting extraterrestrials is therefore being directly assisted and monitored by covert authorities in the US and elsewhere where such assimilation facilities are located.

JP also travels to a subterranean castle in an icy region of North America with a massive stash of gold plates that are being loaded onto shipping containers for transport worldwide. This transfer may be part of an Earth Alliance project to convert our planet from FIAT to gold-backed currencies. Such a conversion process will go a long way in freeing nations from the debt slavery created by Deep State-controlled central banking systems.

In the now familiar underground Ant kingdom, JP learns that a sleeping giant has awakened to become the Ant People's new king. This king will lead his people to a new realm and human refugees who long ago sought shelter in the Ant kingdom will be left in charge of the habitat. JP describes how he was able to reconnect with the Ant People despite them disappearing from another known underground habitat and he is told he will visit them in their new vibrational realm. While he did not meet their new king, he is confident this will happen in a future mission.

In preparation for an operation to Saturn, JP undergoes a series of mysterious testing procedures. Afterwards, a senior officer tells him it is likely he has already completed the Saturn mission despite him having no memory of doing so. This information is deeply disturbing to JP because of its implications, and he realizes that he has performed more covert missions than the 35 he has reported to me during his military career. JP comes to understand that he will be able to remember more at a later period, when certain conditions are right.

Yet another mission takes JP to a massive underground spaceport somewhere in Alabama, where 30 top military pilots from the US and other nations are training to fly a Nordic-designed saucer-shaped craft. The craft have highly advanced technologies

PREFACE

on board, and it's speculated that such training is a reward for countries signing onto the US-led Artemis Accords. In a roundtable discussion between JP, Elena Danaan, and myself, more details are shared about the Nordics taking over the space arks, the awakened sleeping giant who has taken the Ant People to a new realm, and JP's trip to Saturn. The Saturn mission was particularly dangerous because it involved a formerly *forbidden zone* under Anunnaki control, which only recently was transferred over to the Earth Alliance. Finally, in a moving exchange both Danaan and JP share the physical health toll they have experienced during their respective journeys to outer space or the Inner Earth as a result of their exposure to different densities and electromagnetic fields.

New readers of this series will find it helpful to know more about JP's unusual history. In 2008, JP had his first encounter with Nordic extraterrestrials who activated his ability to interact with ancient technologies such as those found on space arks and in underground civilizations. These extraterrestrials had also made secret technology exchange agreements with the USAF, which included disclosure requirements.

JP enlisted in the US Army in 2019 on the advice of USAF military officers who had encouraged him two years earlier to expose the existence of a secret space program operating out of MacDill Air Force Base in Tampa, Florida. JP subsequently released dozens of photos of highly classified antigravity flying triangles and rectangles, including those he had been taken aboard during military abductions. Upon enlistment, JP gained access to the military's highly classified special access programs and due to the USAF disclosure agreement with the Nordics, he has been allowed to divulge unprecedented secrets. At the time of writing this preface, JP's status with the US Army and the covert USAF military hierarchy is about to undergo a fundamental change. In the final chapter, I will discuss the upcoming change to JP's Army status and what this means for future covert military missions.

Michael Salla, Ph.D.
August 25, 2024

US ARMY INSIDER MISSIONS 3

Chapter 1

Returning the Activation Jewel to the Atlantic Space Ark

On September 11, 2023, JP was part of a joint mission to return an activation jewel taken from the Atlantic Space Ark a few months earlier in July. He described how five Nordic extraterrestrials handed off the jewel to JP's team of four special operators after they had returned it. Four of the Nordics joined JP's team as they descended into the space ark from a donut-shaped naval surface ship. After it had been taken from the space ark, the jewel had been used to activate ancient technologies around the world.

Once again, JP described the deep emotions of sadness and happiness created by the jewel, which affected all joint team members, including the four Nordics. After completing the mission, JP said the ark had begun moving and would soon detach itself from the elevator shaft connecting it to the naval surface ship. JP was told the space ark was now on the move and was expected to rise into space with the approach of a new Atlantic hurricane. The transcript of the interview follows, with grammatical corrections and my commentary.

1

Interview Transcript with Commentary

Key: MS – Michael Salla; JP – Pseudonym for US Army Soldier

MS - I am very pleased to have JP back on Exopolitics Today. He has an update about the Atlantic Space Ark, so welcome JP to Exopolitics Today.

JP - How are you doing, Doc? It's a pleasure to be here on Exopolitics Today. I know this is back-to-back, and this is awesome. We're living in awesome times and I'm getting the green light. So that's really awesome that when we do missions like this back-to-back, I get the green light to tell you. So that's really good, you know.

MS - Well, this is great news. I always enjoy hearing what the latest missions are, and of course, you've done four missions to the Atlantic Space Ark, and in the last one, you talked about a jewel being taken out. You mentioned that there would be another mission where the jewel would be returning, so I guess this is that mission. Is that right?

JP - This is that mission.

MS - So tell us, how did it begin?

JP - It was a total of four of us, right, including the pilot and the co-pilot on UH-60 [Sikorsky UH-60 Black Hawk Helicopter]. We were on a UH-60, and we flew out to the Atlantic, and we landed on the donut-shaped ship that was actually ready to leave its location at this particular time right now. So, we flew, and we were dressed, dressed up in all blacks, black uniforms - a type of OCP [Operational Camouflage Pattern] but black ones. We landed, and we saw five beings. They looked human, but they had slightly bigger eyes. Nordic-like extraterrestrials, but they were dressed in uniform without name tags - no names or anything like that. And they had a glass box with them, with the jewel inside of it. This time, there was no ships around [the donut-shaped Navy ship]. So we landed, and they were already lined up with dogs checking them - just sniffing them around, and they were just in formation.

So, we got out of the helicopter, and the helicopter left fast. It got there, and it moved out fast to another location. There's another pad that is out there that they land on to put on fuel. We got a couple of those pads out there. I know the Navy is really familiar with what I'm talking about - for helicopters to land and to get refueled. There are certain parts of the Atlantic or the Pacific where we have these types of pads that, probably Ospreys and helicopters, can land, and they get fueled by certain kinds of ships that pass by and fuel them.

And they're there just in location, like stationary, in grid locations in order for helicopters or Ospreys or other types of military, whatever needs fuel. So we got there, and it was kind of tense because the Nordic soldiers were not talking. They were really hesitant to give back the jewel. So, you know, they were holding it, a total of five of them, and they said - they pointed down. They did not want to talk, and they pointed out that they wanted to go and [leave] the jewel with us.

So, I did not see them - in what vehicle they got there. And how they got there. So we went to the big room. It's like a little horseshoe room. And we got briefed on what we needed to do. So our mission was to bring back the jewel and put it back into the location before this big massive storm passed by. Hurricane Lee, as you know right now, this big storm is one of the biggest storms that the Atlantic has seen in a while. It's a big storm. I know I've been sharing with you on-and-off [information about] the location of this storm. Now, this storm is getting close to the Atlantic Ark. Our mission was to bring this jewel back to the Ark.

Hurricane Lee was a Category 5 hurricane that passed over the Bermuda Triangle area from September 12-15, 2023, and tracked north parallel to the US East Coast before making landfall in Canada's provinces of New Brunswick and Nova Scotia.

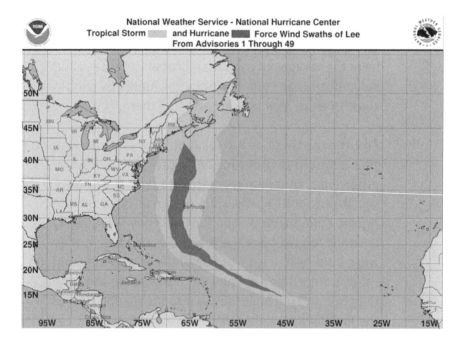

Figure 1. NOAA map tracking Hurricane Lee's Path. NOAA.

MS - Can you just say, JP - today is Wednesday, September 13th [2023], so when was the mission?

JP - This was Monday.

MS - Okay, so we're talking September 11. And this Hurricane Lee is making a beeline for that - for where that donut-shaped ship is located.

JP - Yes.

MS - Okay.

JP - So, it is intentional that this storm is going that way. I don't know what technology we're using, but ... I feel that it is intentional that it's going toward that location. So, we started going down the elevator, and we were looking at four of the soldiers, because one stayed behind on top. So, we decided to go four and four because, you know ... if something hits the fan. You know, we don't know what's going to happen. We have never seen these guys before, you know, but they looked friendly. They were really polite. There was a female one.... She was with them.

She went down as well with them. So, four soldiers and we had four soldiers, and we had our watches with us that measure a heartbeat, where we are at, our location. So, we started going down the elevator to the Ark. And we noticed that the, you know, the highway of water. I don't know what ...

MS - The Gulf Stream.

JP - There are a couple of names there. You've got the Gulf Stream that goes north to south, and then you got the other one that goes left to right. I don't know the name of the other one.

JP is likely referring to the North Atlantic Drift, which, along with the Canary Current and the Equatorial Current, form a giant circular loop with the Gulf Stream see Figure]

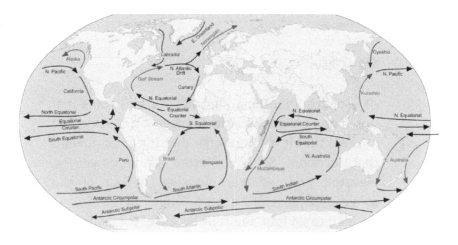

Figure 2. World ocean currents. Wikipedia: Public Domain

But, yeah, we felt it coming down. It was a little bit - the current was a little bit stronger, and we were going side to side until we got to the - this this tube that the elevator, I don't know if I ever got into this with you. This tube that the elevator goes up and down [in]. It's made of a flexible type of metal chain-like technology that moves with the current, but it was moving more than usual. So, you hear the elevator scratching coming down and sometimes scratching every 50 feet.

RETURNING THE ACTIVATION JEWEL TO THE ATLANTIC SPACE ARK

You hear it when it's touching the metal chain wall. You know that the current is strong, and then you're moving side-to-side. You're like, oh crap, oh man. And then this elevator has like two double doors.
And there's one that opens, and then there's a second one that's really thick. It's like two feet thick, that opens, and you go in there. And that's what pressurizes the elevator that takes you down. So that two-foot [thick] door closes first, and then the other one closes on top, and then it takes you down, and it's like a tube, a metal chain tube.

It looks like a massive straw, but [it's] surrounded by metal. Quite interesting, that technology, but it's a lighter metal. I don't know what type of metal, but it flexes. It flexes in and flexes out. So we get to the ark, we open it, and we point it to the other soldiers, the Nordics, to go first. And no, they hesitated to—"No, you guys go first," and they gave us the glass jewel. So when we grabbed the glass jewel, we felt the same sensation of feeling sad.

In his first mission to retrieve the crystal jewel from the Atlantic Space Ark, JP described the feelings of sadness and joy he and the other soldiers felt when holding the jewel and transporting it to the surface.[2] Many wept as they felt the emotional energies the crystal jewel was radiating.

And we were like, "Oh, man, here we go again." Two of the guys that were there with me went on the same mission when we took the jewel. So that one dude was there to put back the jewel - that took out the jewel for the first time.

So, we were going through the chambers, the chambers of the sarcophagus, different types of boxes. We passed the chamber where we saw the writings, different writings. I saw something different that I hadn't seen. I saw another writing that was different. It was ancient as well, but I had never seen it before. I just never looked at that location when we went there [the] last couple of times. So, every time we come down to the Ark, we see something different. You know, you always have to take notes of what you see and report back when you get back and finish the mission. So, we keep walking, you know, and we always have an NCOIC [Non-Commissioned Officer In Charge] in charge, you know, telling us what to do, or, you know, where to follow and how to do things.

And it's really structured, you know, you really can't touch anything. You really have to stay with a low, low voice just in case we hear something else. So, the other soldiers, they kind of were doing the same thing that we were doing. They were following us—to the back of us. So, we kept walking, and we passed the chamber where we saw the fish. The other four Nordic soldiers looked at it, and they were amazed at the floating orb of water that was there with fishes that were glowing.

So, we passed that, and then we went to the chamber where we saw the statue that had the three eyes. [The] statue that looked a little bit oriental, it looked like a Buddha, but not Buddha. It's like Shiva, Buddha; it's a mixture of religions.[3]

It was quite interesting to see again. So, we passed that, and we got to the location where we needed to put the crystal. We went one by one. We did like a half circle, and we told the guy that had the box, we did not know how to open the box. So, the other soldiers, went to the box, and they opened it and they started taking out the jewel. And you could see the tears in their eyes when they were handling this jewel, and everybody started tearing up.

It's a feeling that you get that - it's like when somebody takes something that's precious from you or when you lose a loved one that you really love. That's the feeling you get when you look at this particular jewel. I guess the glass box was a protection. It had a type of protection. I don't know what type of glass it was, but you can still see through it. But it's their technology. So apparently, they came from somewhere else in our solar system, and they activated something else with this jewel, and also, they activated something in Europe, in England.

There's one there. They activated something else, somewhere in South Africa. I do not know where, but I heard Africa. We were told in our briefing before coming down of the activations that this particular jewel has [done]. And then they told us that each of these arks, they have a particular jewel—a similar jewel that activates different things in their region where they're at. But the one in the Atlantic activated something out in our solar system, so they needed it.

So, they brought it back, and my friend, he took the jewel, and he [began] bawling. He was like, "Man, don't look at me". We were like laughing and just bawling at the same time, because we felt like a sense of peace. But it was like a sad - it's a sensation that comes back if you even think about it, you know.

In the Roundtable discussion featuring JP, Elena Danaan, and the author presented in chapter 11, it was explained that the organic consciousness powering the Atlantic Space Ark could be downloaded into the crystal jewel, which could then be used to activate or interface with other arks or ancient technologies. Given the ancient memories of the organic consciousness inhabiting the space ark, it is no surprise that those coming into its proximity would feel powerful emotions, as described by JP.

> So, they took the jewel, and they started putting it back. And it locked in place, and it started glowing, and the ark started freaking shaking. So, I said, "Hey, we got to get the hell out of here." I did not feel that we needed to be there any longer. I said, "We need to leave this jewel alone and get the hell out of here." They agreed. We started going back.
>
> When we started going back, we passed the chamber where the fishes were and the fish were all on the floor with the liquid, just flapping. It looked like the gravity had stopped. Whatever was holding the orb where the fishes were, [had] stopped. But it was coming on and off. You can see the fish floating and then flapping, and then floating, flapping. It was like trying to turn back on, but the ship was shaking. We were looking at our watches, and we were looking at a grid location. Apparently, the ship was moving a couple of feet per minute. So, I said, "Hey, we got to get the hell out of here and get on the elevator before the ship, you know, dislocates. We have to get out of here."

So, we started moving back, and the Nordics, they wanted to stay. So, we're like, "Oh, shit, here we go again." Because, you know, something similar happened to us with the Mexicans, with the Aztec people, that they were dancing and all that.

MS - Yeah, the very first Atlantic Space Ark mission; you said that the Aztecs...

JP - A Kuria Matte, yeah, when they started singing, A Kuria Matte.

JP is here referring to his first mission to the Atlantic Space Ark in January 2022, which was a joint mission between the US and China, as described in *US Army Insider Missions 1*.[4]

The Nordics stayed behind, and our orders were to always stay with them. So I was looking at my watch. Everybody was looking at their watch. The watch was acting kind of weird. It started turning off, and all the electrical stuff that we had started turning off. We were stuck in this chamber for about 40 minutes, and we were there with the Nordics, and they started humming a beautiful song.... It sounded like an Egyptian-Hebrew type of language, and they kept singing it as if they were worshiping something. So, we were hearing that. We were like, "Wow, it sounds beautiful." We're like, "Man, it's crazy, but we still need to get the hell out of here."

So, while they were singing, they were activating the ship. They activated the location where the door opens, and they pointed at us and said, "You guys have to go now." We kind of like heard what they were saying, but they were not moving their lips, so we understood like directly what they were saying telepathically, like talking to us, like, "Move, you guys got to get out right now. We have to stay here. Our mission is to stay here. You guys can go. One of us is up there, and we'll explain to you what's happening." The reason that one [Nordic] soldier stayed behind is to explain to us what was going to be happening. I'm sure that he was explaining already up there to the higher ups what was happening and what they were going to be doing.

So, they actually were like a type of activating the ship in a certain way, that when the storm comes [Hurricane Lee], I think the storm carries this type of energy when the low pressure goes over the ark. I think the storm has a type of electrical [effect on the ark]. I know in the eye, the storm is not that strong but it's really, really hotter in the middle of the eye of the storm. So, that energy that goes around, I don't know, it activates something in the arks.

JP - I don't know how it works, but it's a type of technology that they use for activation. And it kind of makes sense that in the middle of the storm, this ship could leave, and I remember them saying that the donut ship was going to leave the location soon.

The donut ship JP is referring to is a US Navy ocean platform shaped like a donut that has been floating above the Atlantic Space Ark since at least 2016, which is when he was first taken there, but was denied access to the submerged space ark by an Army General.[5]

So, now they're like, okay, so this ship is moving enough, like maybe is going to be fully activated, and will leave with - this Hurricane Lee - to the upper atmosphere and unite with … something that is coming into our solar system right now that everybody's going to see, it's going to unite with it. And we were just, "Okay, we're going to leave you guys here." So we started going towards the elevator, and we started hearing the elevator scratching with the tubular metal chain … that holds the elevator.

And we're like, "Crap, man, hope the higher-ups already know about them staying here." And I told them about my experience with the Aztecs. I said, "Hey, we have to leave them there. If it's for them to activate something." We probably already knew that's why they [higher-ups] called for them [the Nordics] to bring back the jewel. So, it's something that is above our pay grade, so we don't have to stop them or anything. They know, and we have seen that they're activating something on the ship, and we feel peace with them.

They're really peaceful, and they're just doing their job. So, we got on the elevator and we started to close the doors, and I gave one last glance to them. And they just waved and kept humming and singing that beautiful song. And while they were singing, there were activations of different areas of the ship, a type of light that was turning on. It was like a bluish-gold light. It's a beautiful light that was stringing up like in veins and lighting up the ship.

While we were going to the elevator, we were seeing that. Everything was getting lit up, and there were things that we did not see that the light lit up that was really interesting. It was really nice. It was beautiful, beautiful, beautiful. It was a beautiful interior type of light. So, when they were singing we started feeling a type of goodness, peace-like happiness, like joyfulness.

And we kept with that sound, and we started humming it in the elevator, and we were looking at each other like, "What the heck. We're all humming at the same time," like it was contagious, the rhythm. And I'm sure when people hear it, they're gonna keep humming it.

It was beautiful. So, we go up and this elevator, I never felt it going that fast up because I felt really heavy on the way going up. The way the elevator was going up, it was fast. So, we got up, and we saw that one Nordic soldier. He was talking to everybody, and everybody had their mouth, like their jaw open, like "What the heck?"

So, this donut-shaped ship is going to start heading north, towards the middle of the north, to some islands that are close to Europe. I know people, if they put one-and-one together, probably know which islands are these. It's not Bermuda; it's more north.

MS - Canary Islands?

JP - I don't know, can't say, but you guys will put one-and-one and figure it out. I'm sure there's a lot of smart people who look at these videos and they put one-and-one together. You know, I do get the green light to say certain things, but there are certain things that still need to be classified just for the protection of these massive ships and these massive vehicles that we have.

> So, the elevator detaches, right? It stays floating there, like a floater, and the ship leaves it. And I guess whenever it comes back, it knows the location by GPS - that it left behind - and the grid location. But this massive storm is going to be huge. I don't know if people are tracking, but it is massive. It's going to hit right now. It's getting close to the location of the ark, and it's going to activate the ark. That's where we're hearing. That's why they had their jaws down.

JP's mission to the Atlantic Space Ark happened on September 11, and Hurricane Lee began hitting the Bermuda Triangle area with increasing intensity from the next day up to September 15.

> That's what the Nordic soldiers were saying: that the ark was going to be activated. And the helicopter was there waiting for us already. So after the briefing, we told them everything that we saw, and we even told them that there were even more writings that we did not research, and this and that, so everybody was kind of sad because of that.

> But the ark was moving a couple of feet every minute to the deep side of the Atlantic, so activation or not, it was moving towards the middle of the Atlantic to where it's deeper. So, I know to the west side of the ark, there is a city there of the three pyramids and these beings that live there, we haven't communicated with them in a long time since the last interview that we had.

JP is here referring to a complex of three domes with city-like structures inside joined by long glass-like tubes in the Bermuda Triangle that he visited during a mission in August 2022 that was described in *US Army Insider Missions 1*.[6] Important corroboration for what JP saw emerged when Jean Charles Moyen released information about an underwater complex of three domes joined by clear tubes that he witnessed when he was 10 years old on August 15, 1979. Moyen says that he teleported himself to the Bahamas and brought back with him some physical evidence of what he experienced, which he showed to his parents, who kept it for safekeeping.[7]

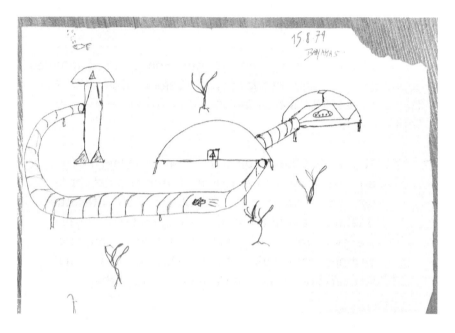

Figure 3. Underwater complex of three domes that Jean Charles Moyen witnessed and drew in 1979.

JP - I know there are other people that communicate or have missions that are more intense. I know there are more people out there, Doc that are going through these types of missions and have even a bigger story. But, like we know, I'm getting the green light to tell you this information, and when this type of information comes out, people get attacked, you know. I'm getting attacked here and there, for me to

> stop talking, for me to stop bringing out this information, but, you know, we got the White Hats that are saying, "Hey, no, you're good, you're protected."

The White Hats JP is referring to are the covert branch of the US Air Force and other military services that are in favor of UFO disclosure. JP was first encouraged to reveal information concerning secret space programs involving antigravity craft in 2017 when he took photos of triangle and rectangle-shaped craft operating out of MacDill Air Force Base. The photos and encounters he had with covert USAF officers encouraging him to take the photos and disclose the information are discussed in *US Army Insider Missions 1*.

> So, you know, there's a battle right now. It's massive with disclosure and people need to know that right now that this battle is really getting fired up with bringing this type of information out.

> JP - You know, there's one thing talking about, "I saw a ship and yeah, we followed it" and all that, but there's other information that they're probably leaving out that they probably visit the ships or they probably visit other places that are in the Inner Earth that they're not allowed to talk to.

> MS - Can you talk about what the jewel from the ark was used to activate in other places? I mean, you said that the other things were activated. I mean, are we talking about other arks or technologies in different places? You mentioned Europe, and I think you mentioned Asia. So yeah, can you talk any more about what you were told about that?

JP - Ancient technology type of ships that are dormant, different types of smaller ships that are protected by certain governments that are really advanced types of portal-like technology. It also activates certain people as well, like, say we bring this jewel to a location where there are probably 300 people, right, out of those 300, probably one person will get activated in a certain way that he could see the past and he could see the future. The jewel has that type of, I could say, energy or power that activates certain individuals to see the past and to see certain events that happened. So, it's like a type of library, but it goes to the quantum realm into your head and it activates visions. So, it's a really interesting technology.

MS - And who was doing the activation? I mean, who was carrying the jewel? Because you say the Nordics brought the jewel to that floating, donut-shaped ship. So were the Nordics doing the activation, or was it the Earth governments?

JP - Well, the same way they came back to bring the jewel and to put it back into the ship. There are Earth governments, you know, we were part of that putting it back into the ship. So, there are Earth governments that also get involved with them. And there's only particular individuals that can go into the ark and there's other individuals that go into other arks that we can't go into. So, I guess they pass it on to the other individuals who are allowed into other arks. They activate it, and then they give it back. You need to really trust people to handle this type of technology. It's really interesting the way it's organized, you know, and it's protected. I'm sure there were a couple of unidentified flying objects flying over the donut ship protecting the situation or inside the ocean, just protecting whatever that we were doing, you know.

MS - You mentioned that the ark was moving, so is this movement just to avoid this hurricane that's coming, or is this movement part of something else?

JP - This movement is part of the activation of this particular ark that is probably gonna lift up through the storm to the atmosphere.

MS - So when you say lift up through the storm to the atmosphere, you're saying as this hurricane is above the location where the ark is, the ark's just going to go straight up. And so effectively no one's going to see it because there's a hurricane above it, but the ark will use that as a cover for what, to go out into space and stay there for a period of time? Is that what you were told?

JP - When we were leaving, right we got on the helicopter, you know, the other people there are getting the briefing. And we looked down to the helicopter Doc, there was a major green glow coming out from the ocean the size of the ark. It was green, a bright dark green light coming out from the bottom, and that's something that you can totally see from space—the satellites can totally pick it up. I'm sure they could pick up the donut-shaped ship as well, and you notice that when we dropped down the balloons, we never made it we never let it go to the Atlantic. Did you notice that?

MS - You mean that Chinese spy balloon that was shot down? Is that what you're talking about?

From January 28 to February 4, 2023, a Chinese spy balloon moved over Alaska, Canada, and the continental US conducting surveillance, which included sensitive nuclear missile sites such as nuclear silos in the Midwest. The spy balloon was eventually shot down by NORAD soon after it entered the Atlantic Ocean. In the following portion of the interview, JP alludes that the flight path of the balloon would have taken it over the location of the submerged Space Ark and the donut-shaped Navy platform above it.

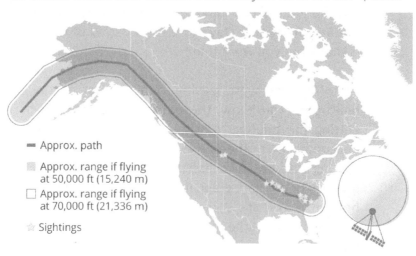

Figure 4. Open Source Intelligence. Statista

JP – Yeah, it never made it out to the Atlantic. There was a reason for that. Yeah, people can put one and one together. I know a lot of people are smart. We were flying up right with the UH-60 [Black Hawk helicopter], and we started seeing the whole area where the donut-shaped ship was glowing a greenish color, and it was coming out from the deep. That means this ship was really activating the waters around it. And we also saw the same thing above the [submerged] city. It was glowing blue, like a bluish, dark, bright blue light coming out from that location. So, it was activating the city in a way, but it was also activating itself and the ocean.

MS - And you're talking about that underwater city with the three domes, right? The one that you described in an earlier mission report. So, you describe that as, even though it's a complex of three domes, you describe that as what, one city or three cities joined together?

JP - It's like three cities joined together, but it has its own government. They operate under the same ruler.

MS - What happens after the ark, if it uses the hurricane to go out into outer space? Once the hurricane leaves, what happens then?

JP - They start activating one by one, I guess. All the different arks start activating, doing the same thing.

MS - Are the arks going to do it in a stealthy way, like this Atlantic Ark? Because you're saying it's going to go out into space using a hurricane as a cover. So, it's like there'll be a stealth maneuver into space. So, is that going to be the same with other arks? That they're all going to do something in a stealthy way to go out into space? And then what are they going to do up there?

JP - This particular ark is gonna meet up with something else that is coming from space. I don't know what it's gonna meet up with, but this week coming up, it's gonna meet up with something in space. And I think a lot of people are gonna talk about it.

MS - OK, so I know more than a year ago now, you said that when the arks start activating or moving to a high mode of activation that they're going to start floating, and people will see them. Is this part of that process or something else?

JP - What we were told is that this particular ark is beginning to activate. And it's going to get lifted up when this particular storm comes through the Atlantic. And it's going to go higher into the atmosphere. And it's going to reunite with something else up there. This particular ark is different from the other arks. It's not the same with other arcs. Possibly, I hope we can get to see this beautiful ark and everybody can see it. We need to figure out where this hurricane is going to go. Where is the pathway of this hurricane? So this this ship is quite big. You know, I'm sure there's going to be a sighting somewhere. Of this particular ship, this particular ark.

And this week, more orbs are going to be seen coming in and out from the oceans. There's a massive activation happening right now in the

> oceans. A lot of things are getting activated in the oceans. There's a lot of activation happening in the oceans. And people are going to see these orbs and see these ships that are like half the size of the arks. But yeah, this particular ark is being activated. And that's where we're told from these Nordic extraterrestrials. And that's what I got the green light to say.

In US Army Insider Missions 2, I discussed JP's revelation from February 2023, that many orbs were being released from space arks and underground spaceports that would be seen around the world.[8] This was confirmed by Dr. Sean Kirkpatrick, then Director of the Pentagon's All-domain Anomaly Resolution Office,' and Harvard University Professor Avi Loeb, who speculated that recent UFO sightings may be orbs originating from alien motherships (aka space arks) in our solar system in a joint March 2023 paper.[9]

> MS - Okay, and so for now, as far as you know, the Nordics are still in this ark. Is there going to be a mission to rescue them, or is it just going to be up to the Nordics to deal with their personnel? ...
>
> JP - Yeah, I think right now it's up to the Nordics because I'm sure they dislocated the elevator, and the way that one Nordic was talking is that they have it handled. That's why everybody had their job, like we got this handled. I guess they got told from a higher-up and they relayed the message to them that they're in charge now. So, it is they that they are in charge right now of that ark. We don't have nothing to do with that ark right now, and I'm sure they're going to do the same thing with the Pacific Ark.

MS - OK, so that was the Atlantic Ark in that area of the Bermuda Triangle. Now you mentioned the Pacific Ark, which is in the area of the Dragon's Triangle, South of Japan to the east of Taiwan. So, do you know anything about that one? I mean, is that activating?

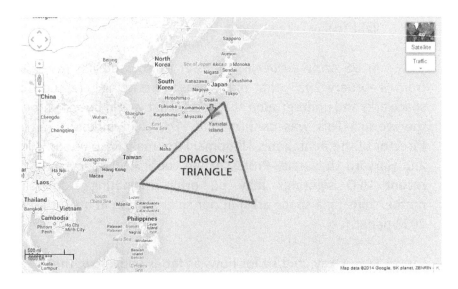

Figure 5. Dragon's Triangle, aka Devil's Triangle. Source: Google Maps

JP - I'm sure it was probably activating at the same time as this particular ark, but I don't think that one is going to the atmosphere. The Atlantic Ark is going to come up from there, and it's going to meet up with something that is in our solar system that is coming in.

MS - It sounds as though the Atlantic Ark, I mean, you've mentioned before that it's the biggest of all the arks, so it is seemingly playing some kind of pivotal role in how all the rest of the arks are going to behave in response to something coming into our solar system. Is this something new that's arrived in

> our solar system or something that's arriving near the Earth that's been in our solar system?
>
> JP - It is going to be near the Earth, but it has arrived in our solar system before. Everybody's familiar with this particular object that is coming in, I didn't get enough of the green light to talk about this particular object. But I'm sure people are going to talk about it soon.
>
> MS - Okay, well, you know we've had Elena Denan and a few others talk about fleets of craft arriving recently in our solar system within the last two years. Some of those have been moving towards the Earth, so specifically, it's one of those ships.

Elena Danaan first discussed the arrival of fleets of Seeder ships, aka Intergalactic Confederation, arriving in mid-2021.[10] This arrival happened at the same time. JP was first forewarned by Nordic extraterrestrials about the arrival of the visitors, and him joining a multinational mission led by US Space Command to meet and greet the new visitors around Jupiter's moon Ganymede.[11]

> JP - They're, actually, going to in a vicinity of Ceres and the Asteroid Belt heading close to Mars that particular route right now. But yes, they're really getting close to Earth, these objects. And they're huge. They're really huge ...
>
> MS - Now, last time that you went into the Atlantic Space Ark, you had missing time. You were there for a couple of hours, or you thought you were there for a couple of hours, but it turned out you were there for two days, and you had flashbacks or visions of extraterrestrials down there. So, did you get any

more information, any more clarity about who they were that you saw down there?

The incident referred to here happened during JP's 4th mission to the Atlantic Space Ark, which occurred in July 2023. It involved the initial extraction of the crystal jewel from the Atlantic Space Ark and the powerful emotional effect it had on all the soldiers participating in the mission and their discovery of two days of missing time.[12]

> JP - No, I know that these soldiers that came, they look really similar to those people that I saw. Really similar to what I saw. That's one of the reasons we felt kind of at peace, because I know my friend who went with us to pick up the jewel felt it too. He was there with me, and he felt the same way. He's like, these people are familiar, these people are nice, you know, and the way they were tearing up as well when they were handling the jewel. We knew that, yeah, they're really peaceful. We feel peace with them.
>
> MS - Also, so those Nordics did tear up when they were handling the jewels?
>
> JP - Oh, yeah, they totally teared up. They were tearing up as well as us.
>
> MS - That's interesting because one would have thought when the jewel was taken out originally, and people you described tearing up and people crying and having that emotion, that returning the jewel would have the opposite effect, but it's interesting that it was the same thing.

JP - Yeah. This particular jewel holds everything to it, you know. It grabs the sadness of everything that it connects to. The reason I think they took it to activate these other places, was just to take out the negative. I think it takes out the negative part of everything, and it just keeps it to itself. And just by looking at it, you just feel like crying. It's crazy. And certain people, they can see, if they touch it, they can see the past and they can see major catastrophes that happened before. And it just flashes. Like if you were blinking really fast and you see these flashes of destruction, but not necessarily of our planet, but of other particular planets. Yeah, it's really interesting, Doc, what's going to happen. I know a lot of people are going to have a lot of questions.

MS - So of your team, of the four of you that went down there, the soldiers, that you were part of that squad, did any of you guys have any kind of visions when you were near that jewel?

JP - Those were the flashbacks I was talking to you about, Doc, that loss of time. We all touched it in some way, the jewel, you know, the handling of it. So you get these flashbacks, and it's millions of years old, you know, older than you can imagine.

MS - Okay, well, sorry ... we've been going for some time. So anything you want to say about what you think is next when it comes to the Atlantic Space Ark, or the arks in general?

JP - Okay, well, I was told that this ark was going to be activated. I know there's going to be, there's going to be a mission to verify, but we don't know when that is going to be, to verify, to see, if the arc is still there or to see where the location of the ark is because we do have, we did leave stuff in the ark and we left ... stuff there that certain people wanted us to leave that, you know, I don't have a particular green light yet to tell you what we left there, but it's certain stuff that we left there about our history. Just in case something did happen ... they would know about our civilization or know about how we lived and all that. So we, we did leave stuff in there, in the ark.

JP is here saying that the Atlantic Space Ark now carries records of our present Western civilization, which accompanies records of previous civilizations on Earth that JP has witnessed in his previous missions to the Ark. It appears that this is a common practice among visitors to the Atlantic Space Ark, similar to etching one's name to an old oak tree. Presumably, as JP next acknowledges, what was left can also be electronically tracked, allowing the US and other national militaries to track the space ark as it travels under the ocean and/or into outer space.

> JP - So we're going to know where this Ark is at. If it does go up into the atmosphere, we're going to know the location. It is moving more into the middle of the Atlantic. We don't know the location—if it's moving to another type of [underwater] city to activate a certain city that is in the middle of the Atlantic, that could be a possibility. But what we were told from the Nordic on top is that he is going to go meet up with something. Who knows, maybe it might come back, and all will see it. It will be really exciting if that happens.... What I want to leave with the people, if you guys see arks or a particular ship, is that we're going to have more sightings. That is

> the reason every outlet on TV is talking about it. There's something that's going to happen.
>
> Something is going to occur. That's going to shock a lot of people and a lot of religions all around. People are going to be shocked when they see this for the first time and they see it as clear as day, but we also got this Project Blue Beam that we have to be careful of, you know. So, it can be a mixture of Blue Beam and real, real things happening at the same time, just to confuse people.

JP is here referring to a false flag event involving UFOs that has been called Project Blue Beam due to its anticipated use of holographic and other advanced technologies to simulate an alien attack of some kind. UFO researchers have evidence of Deep State plans to stage a Blue Beam type alien event going all the way back to the infamous 1967 *Report From Iron Mountain*.[13] JP is suggesting that such a staged event could be launched by the Deep State at the same time as genuine UFOs involving space arks, orbs, or Seeder motherships began appearing in the sky, thereby confusing everybody about the true origins of the witnessed craft.

> JP -So you just have the heart and mind and happy thoughts just to discern what is real and what is not. And that's what I want to leave with the audience today, prepare yourself. You know, we're getting ready to ride a nice storm and there's a battle right now for disclosure, and we just have to buckle up and, and ride the tide, you know, Doc.
>
> MS - Yes, thank you. That's good advice. So, is there anything you want to say in finishing to our Portuguese and Spanish-speaking friends?

At this point, JP spoke in Spanish and Portuguese, and then paraphrased what he had just said to the English-speaking audience.

> JP - So, Doc, basically what I said is I'm really happy because every country that has a story is part of this. It's not just the United States. You know, every country has its own type of story of what's happening here. And I know that their governments are working [on it]. Maybe they want to keep it classified or keep it between themselves, but we know that most of the governments of the world right now are passing through these types of situations with ark activations or visiting caves that have not been visited for thousands of years, and finding drawings and finding different things that are connecting the dots for different countries.
>
> And that's how different countries are getting involved right now in space. And it's really exciting times, Doc. So basically, that's what I told the Spanish and Portuguese audience to be ready and to keep listening to this information. If they do have a type of information, they may connect with Michael Sala, exopolitics.org. ... If they work with me, they connect with exoolitics and Dr. Michael Salla. Basically, I'm telling also the audience don't be scared, you know. If you're going through these different situations or these different experiences throughout your government, don't be scared to connect with exopolitics with you, Doc. That's how I connected with you for the first time [in 2008], as you remember, with my experiences, and the rest is history. So, I know more people are going through similar things.

MS - Well, thank you, JP. I wish you Godspeed in your continued service and look forward to future updates.

JP - Roger. Appreciate it. And I hope everybody shares love, a lot of love, and happy thoughts. It goes a long way. Do good things to others. If you see somebody on the side of the road with a flat tire, stop and help them. Help them. Somebody needs help with something. Just help them. Always share love. That will activate certain things in your life. I'll just leave it as that. I love you guys.
MS - Thank you, JP.

After Hurricane Lee passed over the Bermuda Triangle area where the Space Ark was located, JP learned that it had not flown into space after all. Instead, the ark had moved underwater to a more centralized region of the Atlantic Ocean. The Nordics were now controlling the ark and activating its different systems thanks to the return of the crystal jewel containing the plasma consciousness animating the ark. The Nordics believed that relocating the ark further into the Atlantic Ocean would reduce the influence of the US over the ark and encourage greater international cooperation. The reasons and process by which the Nordic took control over the Atlantic Space Ark will be addressed in Chapters 3 and 4.

Chapter 2

Mission to Nordic Inner Earth Civilization for Life Extension Technology

On October 5, 2023, JP was part of a team of four soldiers led by a medical officer who traveled to a deep underground civilization located under a major military base. Once at the Inner Earth location, JP described meeting a group of Nordic-looking humans who greeted them and gave them access to an ancient library. The Nordic provided information on life extension technology that was accepted by the medical officer.

In this update, JP describes traveling to a large military base where he and the team entered a small building with an old elevator that traveled 16 floors down. Once at the bottom level, they exited and walked a corridor with many rooms filled with personnel working on computer terminals. They witnessed several small flying saucer-shaped craft. They then entered another larger elevator that appeared to have been built by another civilization that rapidly descended deep into the Earth's interior. Upon exiting, they were met by the Nordic-looking humans who spoke with a refined English accent and also telepathically communicated with them as they completed their mission. The transcript of the interview follows, with grammatical corrections and my commentary.[14]

Interview Transcript with Commentary

MS - Well, I'm with JP, who has returned recently from a mission he performed on October 5, and he is here with Exopolitics Today to tell us about his mission. Thank you, JP, for coming back on the show.

JP - How are you doing, Doc? It's a privilege to be here every time to bring this information. I got the green light, so let's do this. Really excited for this one. It was an interesting mission. Yeah, if you want me to start, I'll start.

MS - Yeah, please just take it from the beginning. Just tell us what happened. And if I have questions, I'll just ask them as we go along.

JP - Roger. In the morning, I got a text, and it gave me numbers of grid coordinates. When we get those types of messages, we know what's going down. So, I was in civilian attire, and I got a text saying, "Hey, go to this location." So, I woke up at 4:30 in the morning and got ready. I was wearing all black. I drove my POV, my personal operational vehicle, to this location, and when I got to the location, it was probably 40 minutes away from where I live. It was still inside the base, if you know what I'm talking about. This is a huge base. So we went to this location, and there was a total of four cars and four POVs. So they also got the message: four soldiers.

One of them was a doctor, a cool guy of really high rank. Could not tell you the rank, and then people can start narrowing it down. You know how that goes. So yeah, he was a high-ranking officer. So, he was with us. We went to this location where there was like a little cabin. It was made of concrete. It was probably a 20 by 20 [foot] cabin, and it was like 12 feet high, and the top part of the roof is actually painted the same color as the trees. By plane or by drone, it looks like a bush. It doesn't look like a cabin. It's painted green army style. From above, it looks like a bush. So we got in, and there was a big door, the middle door. It was four of us.

We entered. We were in civilian attire. I had my sneakers on. We had black on blacks. So we entered, and I said, "Hey, so everybody got the text message," and one of the guys was like, "Yeah, man, they told us to come here." And I said, "I know what's probably going down too." They were talking about this type of mission before going to a place to pick up some information and some intel on stuff that was happening.

So, we entered through this cabin, and there were two metal doors, and it opened. He put his hand on the screen, and it read his hand, and the door opened. I'm talking about the doctor. So, he knew this place. He had probably been here more than once, and we had all entered. It was five by five [feet].

It was a tight spot, but it was an elevator. We looked at the elevator, and it had 16 different floors. I was like, "Oh shit, man, we're going under it. We're going to the underworld." We started just laughing and talking about it. "All right, we're going down. All right, let's go." So, he pressed 16. Bottom one, number 16, and the elevator started going down. Lights were flickering while it was going down. It took us probably four minutes to go down to the 16th floor.

So, once we got to the 16th floor, the doctor went out first. He showed his badge. There were two soldiers dressed in uniform. They did not have name badges. They did not have anything [insignia or name badges] on .. their uniform, and they were asking for our badges. We showed them the badges, and they scanned them. They also scanned our hands, and they also scanned our eyes. I had never had my eye scanned like that before, but I assumed I was already in the system because he said, "Okay, you're good to go." So the other guy the last guy he got scanned to, and he was good to go.

So, all four of us passed, and there were two other soldiers who said, "Okay, come over." It was beautiful. It's like a little complex of offices and windows and a lot of computers. There were people sitting down working in each room. They had like four screens... When I was walking, I was counting probably 18 side-by-side rooms. Kept walking and seeing rooms with computers and people working on computers. It was probably another organization that I won't say, but we probably have an idea. We know who it is.

MS - Okay, I have a few questions about what you've said so far. You know, first of all, the building itself, it's 20 feet by 20 feet, so you know, not all that big. But this elevator that the four of you got into, I mean, you said it was five by five feet. That's tiny. I mean, that is tiny.

JP - Yeah, it is tiny.

MS - Four minutes to travel 16 floors.... We're talking about a very old-style elevator. I mean, I've been in the old elevators.

JP - Yeah, it was an old-school elevator .... The lights were flickering when we were going down.

MS - That's 16 floors in four minutes. I mean, that's slow.

JP - Yeah. It was slow. It was not fast. So, we got there, but the other elevator that we got on was a different type of elevator, but we're going to get there.

MS - So when you're passing all those rooms ... you saw about 18 rooms, and there were computers in there, so you know what were the computers like? Were they modern computers, or were they old?

JP - Oh, everything down there in those rooms was modern. We had guys with gloves that were moving. I don't know what they were doing with their left hands, but they were moving with one hand and typing with the other hand. So, I don't know if it was controlling a type of drone or I don't know what it was, but I know it was high tech stuff they had there. And we were all just passing by, and there was a line. There was a white line that we had to follow. And keep going straight; we didn't stop. The two soldiers there was a soldier in front and a soldier in the back of the fourth guy. And they just told us to keep walking straight. You know, we don't try to look side by side, but you know, in the corner of your eyes, you can clearly see what's going on.

After we pass these rooms, we enter another door. It looked like a little hangar. And there were types of flying saucer-looking ships parked, but they were really small, probably the size of a Corvette. We saw a total of three, and they were all the way on the ground. They were not floating. They were all the way on the ground. So, we saw a total of three, and we passed this type of hangar and it was dimmer than the hallway.

We had a bluish dim light, and the dim light was pointing to these flying saucer-looking ships. In the middle ship, you can clearly see an opening, and the door was small. So, I don't know who was in charge there or which entities were involved, but we did see two flying saucer-looking ships, the size of Corvettes, really small. They were parked right next to each other, and the middle one had the door open. So we passed the hangar-type of place, but we kept looking straight.

MS - How tall were those kinds of saucer-shaped crafts?

JP - The size of a Corvette, I walk by them and it's by my elbow.

MS - Okay, so either a very small person or a being could go in them.

JP - Yeah, totally. I thought it would be like a type of drone type of ship, but when I saw the little door, it made me think about probably another being that is also visiting the area.... The day before, we saw a couple of lights coming in and out from the base. It was quite interesting because they were in formation or like a triangle, but it wasn't a TR-3B or anything like that. It was separate ships, and they were coming in [formation of] threes in and out. It may have been helicopters, but we didn't hear helicopters. They came in formations of three, so it was kind of cool. That was a day prior that we saw that. And then I see these three flying saucer-looking ships that can probably be drones or driven by something else.

So, we passed this hangar, and we go to this other room, and there are soldiers, there are people dressed in black, the same way that we are and they're scanning us again. So, we were in line to get scanned again, and behind them were really beautiful metal doors. These doors were huge. I say they were probably 12 feet high and probably 18 to 16 feet wide. So, this was a big door. So, they scan us. They scan our eyes again. They scan our access card. They scan our hands, and they pass us through a laser type of light that we have to pass by. The security was a little bit higher than usual.

I don't remember seeing this much security when I was passing by. Well, usually they're like, "Oh yeah, just go," you know, but the security here was a little bit higher than usual. After we left that room, the other two soldiers left, and they left us with these other guys that were here in the room where the door was behind them, and we entered the door after we passed the security section. We entered the door, and it was another elevator, but this elevator was a little bit bigger than the one we were coming down. It was way bigger. We could fit [inside] really well, and it looked really high-tech.

It had four lights in the corner, similar to the elevators that we used to go down to the ark in the Atlantic, but this was a fixed elevator that's been here for a while. It's not our technology. This type of elevator was built by somebody else. It's not our elevator, so it's a different technology that they use to go up and down.

MS - What made you think that it was built by somebody else?

JP - The way that it was built was more oval in shape. It was not squarish. And the numbers that were on the elevators were not the numbers that we regularly see. It was like …a Morse code type of numbering thing that they had. And it looked like a different language on the top part of it where the numbers were or where the letters or numbers were. I don't think it was made by us. By the way, I'm going to tell you how it went down. So they told us, the guy said, "Are you ready?" And I'm like, "What do you mean? Are you ready?" And we're like, everybody was looking at each other. "Yeah. What do you mean?" So, he pressed a code. It's not pressing a floor number here. Here, it's by code, by different codes that you put in. That's where you go; I guess that's how it works.

So, he put in some code, and the elevator started going down. Boom. And then I felt like the butterflies in your stomach when something goes down, like really fast. And then we were feeling really light. So one of the soldiers threw his pen up, and the pen started floating down. I'm like, "Oh shit, we're really going down fast." So, I took out my access card, and I threw it up and it was going down in slow motion.

So, we were going really, really fast, really fast. It was crazy fast down this elevator, and everybody was holding on. The elevator was a little bit shaking, but then it kept going smoother until we heard a humming noise. It was like humming constantly, and we're all holding on, and our eyes are [wide] open. We're like, "Where the hell are we going? Oh, this shit's crazy." The doctor said he's done this a couple of times. And he was just chilling and rocking back and forth. It probably took us six minutes. It was a long way down six minutes. So, if the first elevator was like four minutes, this second elevator was six minutes, but the way that it was going so fast down, it was really fast.

MS - While you were going down in that elevator, what were you holding on to? Were you holding on to some kind of support?

JP - It had rails that went all the way around, but it looked like a ring, a ring with a light in the middle, a bluish light that we were holding on to. So, it was oval inside and the floor was slightly also oval in, like if it was going to a speed in the outer part of it, it would go like really fast. It's not squared out. It's like an oval ramp type. It was weird, but … we were comfortable. So, it started slowing down, and the Doc started getting really excited. So, I look at the other guy and say, "Have you been here before?" And he's like, "No, man, I haven't been here before." And I asked the other guy, "Have you been here before?" He's like, "Yeah, I've been here once, man, this is crazy. It's going to blow your mind." I'm like, "damn, bro, this is crazy. All right, cool." So, the door opened and it was a big, big cave system, like huge. So, the door opens, and we walked two, three meters out, and you see a big cavernous system…. It was huge. It was beautiful, and it had vegetation. It had a river that went through it. It was beautiful. You can smell the vegetation. It had a citrus-type smell, and I was stuffed when I started coming down. And once I hit the air of this place that we entered, my stuffiness went away. My eyes stopped itching.

MS - What about lighting? How could you see down there?

JP - It was like bluish lighting that was on the walls, beautiful, beautiful, bright blue, like streaks of lights, like on the cave systems, but it had buildings, but the buildings were in graded like into the cave systems. And you could clearly see glass. They had glass in their windows and all that, but it was a type of metallic-looking as if it were tinted windows and there were oval type windows. So, the structure of the buildings that were there was quite interesting. And then we kept walking, and we met up with this group that was dressed in white, and they were Nordic-looking. They had long white hair and a lot of jewelry on them. And they said, "Follow us, follow us." They talked English really well, … [an accent] more from England, but they talked it really calmly, like, "Hey, come with us." So we started going with them…. So it was a total of four of us. The other soldiers that took us down they went back to the elevator, and they just shot back up. So, we started walking with this guy, and we saw a huge, huge dog.

It was the type of labrador mixed in with probably a mastiff or something like that, but it was huge.

I'm like six foot [1.83 meters] ... His head was like 5' 10" [high]; it was a huge dog. And the dog was also white; his fur was white, and he had whitish, beautiful eyes. You can see a couple of these types of dogs going around and up and down with different people dressed in white. They were dressed in linen, beautiful, beautiful clothes. And they brought us to this building that was ingrained into the cave system. We went in there, and the doctor said, "Hey, wait here!" So, he went in with the guy, and he came out with these beautiful, beautiful metallic books, and when you lift them up ... it was ingrained looking, like 3D. So, you can see different types of lettering on it.

It was something similar to what I saw in the ark. Depending on how you put the lighting on it ... the shadow shows, you can see different types of words. So, it was quite interesting. The books that we were receiving were taken back upstairs. He also gave us the type of medicine that they were working on. They were working on this type of medicine for longevity ... to help us live longer lives. So, they were sharing this with our government. So, I guess they know the [genetic] code for living long periods of time. And I was looking at the buildings, and I could see letters on ... the differently lighted buildings ... similar to Sumerian. So, it was a really beautiful city. And you can see really far, but we could not go farther, but it was a beautiful place, and we received these books that we were going to take back upstairs, back to this place that we were supposed to bring it to.

> MS - So how do you know that the books were about age extension? I mean, did the doctor talk about that? Did the Nordics talk about that?
>
> JP - No, the books were about the history and how they got there. So, I guess they want to share the history of how they got there. I know the book says that the Ant people helped them out to live the way that they're living because they're beautiful-looking people.

JP's reference to the Ant People helping this human-looking race establish a civilization in the Inner Earth supports the idea that insectoid presence on Earth predates humans. In the *Sasquatch Message to Humanity*, Sunbow Truebrother claims that intelligent life on Earth has developed as a result of different extraterrestrial genetic experiments on indigenous species over hundreds of millions of years. The first intelligent species were the ocean-dwelling Aquatics, who were followed by the surface-dwelling Insectoids or Ant People. Next came the Reptilians, followed by the Avians or Bird-people, and next came the Sasquatch, and finally, humanity was seeded on Earth.[15]

> They have blue eyes, Nordics. Nordic beings, totally Nordic [looking]. They had a lot of jewelry on, Doc, beautiful jewelry. They had beautiful bracelets and beautiful necklaces. They had linen clothes, and they looked really peaceful. They're really organized. I probably saw more than 80 people dressed the same and acting the same way, as if they were all connected. So, in the distance, I could see the lighting getting brighter in the distance of the cavern.

It was a huge cavern. In the distance, you could see the lighting getting even brighter. So, I'm sure if we would have walked a little bit more, we could have seen more stuff there. But I will never forget the smell. The smell is so beautiful, like a rose citrus smell. It was like bringing healing into my body by only smelling this. Beautiful, beautiful, beautiful spot. Man, I totally wish people could just visit and see these places, [they are] so beautiful. And I'm sure this is really deep inside, Doc. This is really deep inside our Earth, this place where I went.

MS - What about the buildings? I mean, you said that they were engraved or embedded in the walls. So was it like ... the Anasazi kind of buildings where you can see buildings actually [built] in the walls. You walk in, and there's the building.

Figure 6. Mesa Verde - Pueblo land. By Gleb Tarassenko

JP - It was like it was carved out from the cavern itself. It was hollowed out and carved out, but it was white. It looked like a bone color white, like a pearl white. I guess they have a way of shining it up or, I don't know, and everything is oval-looking. The windows are oval-looking, everything is oval looking, everything is round looking, and that was quite interesting because there's a lot of liquid that falls down through the cavern systems, and it just goes around the building, and it's just beautiful the way they live, amazing the way they live.

MS - So, you know, what else did you see in the cavern? I mean, if the walls of the cavern ... had all these buildings embedded into them, what was inside of the cavern? Was it like rivers, like vegetation? ...

JP - When you first enter this cavern, you see a lot of vegetation. The smell of citrus. They look like flowers, but they're huge. They're orange and yellow in the middle, but they're closed, and they peel off like bananas. They look like bananas when you first see them. And the smell of citrus that comes out from them is beautiful. Different plants. I haven't seen these types of plants. It's more Amazonian-style weather. It's really dense with fog in the distance. You can see parts with different fogginess coming down from it. Beautiful, beautiful, beautiful. They remind me of one of the missions I had with to a dome to a dome system that we went to. I think I told you this probably, like what, 11 years ago or probably 12 years ago.

MS - Yeah, that was, I think, 2015. You went to that dome or that habitat, flying habitat or an ark. You know, so we're talking, yeah, seven, seven years ago, with all the vegetation and so with this, this civilization, these Nordics, did they strike you as like a high-tech civilization or more like, like an Inner Earth, everything, you know, is organic with the environment.

JP is here describing several incidents where he was taken to large dome-like structures in unknown locations where he encountered vegetation and animals that appeared extinct. The first incident happened on October 19, 2015, and he describes being taken to a large dome-shaped "ark" by an orb/sphere most likely controlled by the Nordics he works with. A second visit to a dome-like happened on October 2, 2017, but this time, he was taken there by a triangle-shaped antigravity craft suggesting it belonged to the USAF secret space program. Both incidents are described in detail in *US Army Insider Missions 1*.[16]

JP – [Regarding] Inner Earth and high-tech civilization, totally high-tech. I didn't see any phones or anything like that, but I know they connect with each other telepathically. By the way, they were nodding their heads and talking to us. I even heard the guy talking to me in my head, and I did not see his lips move. I was like, like, "Hey, like you're getting into me." I'm sure they connect with each other. I think they're past the development that we haven't gotten yet. I'm sure that they're high-tech as well. I'm sure they go out from where they live.

MS - Did you see any kind of transportation vehicles? I mean, how did they move around in the cavern?

JP - I did not see a transportation vehicle. I saw them only walking. Maybe if I would have walked a little bit further in, I would have probably seen flying crafts and all that. But remember, on the floor before, I saw three flying saucer ships that were there. So, it couldn't have been from them because they're big. They're like my size people. Yeah, I'm sure they have a type of transportation, but I did not see any type of transportation.

MS - While you and the other members of that small team, four of you, were there, did you see the Nordics or hear the Nordics actually talk, or was it always telepathic communication?

JP - We heard them talk softly, but sometimes we hear them, like, through our heads. We can hear them, as if you're talking to yourself, without moving your lips. That voice that you hear is not as loud and it's not as crisp, like I'm talking right now into the mike. It's not like this. It's more like if you're talking to your subconscious. When you talk to your subconscious, you know, when you're about to do something, and you talk to yourself, that's the voice that you hear. So that's how they sound when they communicate telepathically with you. It sounds like you are constantly talking to yourself, but it's not. And it's actually your own voice, but you know, it's not your own voice. It's not you controlling it. They are controlling it. So, it's like that. It's quite interesting.

MS - So when you went down there first of all, and you saw these Nordics or these Inner Earth beings, you said they spoke to you very quietly in a calm way and it sounded like an English accent. So, is it like that in your head, telepathic, or were you hearing it through your ears?

JP - I was hearing it through my ears, but they also talked in some different language that I did not know ... It sounds more like Catalan. I was doing the research, and it sounds a lot like a mixture of Spanish, Portuguese, English, and French. Similar to what is Catalan, that's the closest language I can relate it to. It's a Catalonian language in Spain.

MS - Okay, so the mission objective was for that medical doctor, that kind of high-ranking medical doctor ...

JP - To receive these books and to receive the technology of longevity.

MS - OK. And did you hear him talking to the Nordics about the content of the books? Did he not say anything? I mean, or did he just open the books and read them?

JP - We were close enough, we were hearing what they were saying and they were talking about the Ant People, how they helped them build that civilization where they were at, yeah, and how they also are all over the world. These same people, they got other cities and other places around the world that are underground, but it's like type of different races. Same way that is on Earth, you know, how we got people from Africa, people from India, people from Asia, you know, that look different. It's the same thing underground. They also have different types of people that they also work together with.

MS - OK, so it sounds like the Ant civilization that they are one of the oldest, if not the oldest, civilizations that live underground and that they've helped others like these Nordics. Presumably, they were once a surface civilization that escaped into the interior to escape some kind of surface calamity. Would that be what happened?

JP - Yeah. Totally what these books were depicting and the history of these people. These people have been down there for thousands of years. I don't know how long they've been down there, but their eyes are reddish-white. They are totally Nordic, blonde, have long hair, and are beautiful people. So, they started walking us out and the linen, their clothes are really expensive looking, like really beautiful, really beautiful. And when they were walking us out, that's when I could hear them telepathically talking to us.

"Oh, I hope you like the experience. We will see you again, and soon, we will also be sharing upstairs with you guys." They say upstairs. "We will surely be upstairs with you guys in the far future." So, I guess these books and these types of medical stuff that they gave us to help us understand what longevity is and all that. I was also talking to this doctor, and we were talking about how I'd seen this doctor before. So, I talked to him before, and they were doing research on transferring consciousness from one place to another. And I asked him, "Hey, what do you mean? How can that be possible? Oh, yeah, you got to get into quantum entanglement." So, he gave me an example, an organ donor, right? When they give a heart to another person, did you know that the heart has 40,000 neurons?

> So, you're basically transferring the memory from one person to another person. The person that has the heart transplant is going to feel and want to eat like the person that had it before. So that's the type of consciousness. So, all they need to do now is transfer the consciousness. But I'm sure that that is possible with the technology that we have right now. And it's probably 100 years more advanced than what the civilian sector has. The way he explained it to me really opened my mind. It's true. Like if you do a heart transplant, you're transporting 40,000 neurons to another person. And that other person is going to have the same feeling that the person who's deceased had. So, it's something similar to that type of stuff.

In 1991, Dr. JJ Andrew Armour was the first to introduce the "heart brain" concept and how the heart has its own neurons, which communicate with the brain. Subsequent scientific studies have confirmed that the heart has its own neurons that are part of a sophisticated communication system.[17]

> JP - You know, I always talk to them here and there about certain stuff and all that, but that's when we're back at that base. We started heading back, and you could hear them in the background talking. We got back to the elevator, and we had to scan again, our hands, our eyes, everything going back, scanning. We have to re-scan everything back together at the elevator. We got to the top floor. We scanned again as if we were coming from another nation or something like that as if we were going through the airport, we scan again. They make us lift up our hands. We pass through the laser, and we scan again. And we have to follow the stupid white line again, back to the elevator, past these guys, past

the hangar, past these guys with the computers. We went back to the elevator, and we went back to the slow elevator back upstairs, us four by ourselves. Yeah, everything took around four to five hours. It was a quickie, but it was really beautiful and a really good experience. And I got the green light to tell you guys.

MS -So that officer, that medical officer, I assume he just took the books or whatever was given to you guys. He just took them with him, got in his car, and drove off?

JP - No, he said drive safe. Okay, drive safe. All right, cool. That's it. We left.

MS – So, there was one officer, and the three of you were just enlisted personnel.

JP - Roger.

MS - OK, and you've got a green light to share this with me and with the public.

JP - Yeah, I got the green light from the guy that always talked to me in the base, and he's actually green-lighting a lot of other people, so I'm quite excited about that.

MS - Oh, can you say anything more about that? His green-lighting other people?

JP - To talk more about their experiences, and to talk more about what's happening.

MS - Are these people that are going to go through the official process, kind of like the whistleblowers, like David Grusch? Is that what we're talking about?

JP – Yes, some of them will go through that process, and some of them will go through the process, like I'm going through, like through you, or the process through the news, or there's going to be different processes of people coming out.

MS - Okay, so some more people are being given the green light to talk about actual missions, like what you've been doing, encountering ancient civilizations and technologies.

JP - It's gonna be more out there, yeah. A lot of people, it's gonna be more out there, and yeah, it's really sad what's happening around the world right now.

MS - I wanted to ask you about that. I mean, you know, you've talked about the arks activating at some point and that there was, you know, that the Atlantic Ark, after your fifth mission, moved kind of more into the center of the Atlantic. Originally, you said that it was going to leave the planet with Hurricane Lee when that came, but later on, you told me that it didn't actually leave and go into space. It actually went more into the interior or more into the center of the Atlantic. So, you know, what's happening with the arks? Are they still on schedule to appear, or are these conflicts like what we're seeing in Israel delaying things?

JP - There's a lot of delay happening right now because of all this. But if it happens, it happens, you know, it's not going to stop something from happening, if it's going to happen. I know it probably sounds like I'm talking in codes, but I can't really mix in any type of conflict with this type of information.

MS - Okay, I understand. All right, so it sounds like...

JP - Yeah, it is. It is sad what's happening around the world, not just in that region [Israel-Palestine]. It's really sad. A lot of negative things are happening that we can't comprehend. If it will happen to us, you know, we can't comprehend that. The arks right now, the Atlantic Ark is run by the Nordics. It's a little bit deeper right now. They're running the [Atlantic] Ark. We're trying to negotiate with them to go back into the [Atlantic] Ark, but they're like hesitating to let us go back to the ark because of the ark in the Pacific. People are in charge of that ark, and they should not have been in charge of that ark, but right now, they are in charge of the ark. I'm talking about the dragon.

The 'Dragon' JP is alluding to here is China, which has projected its military power and authority over the surrounding South China Sea and also into the area where the Pacific Ark is located. China has a long history of hoarding ancient technologies, which it has discovered in its many pyramids and UFO crash retrievals, while doing all it can to uncover, learn, and steal similar discoveries made by the US and other nations. I wrote about China's repository of ancient technologies, UFO crash retrievals, and recent aggressive in gaining access to advanced technologies in *Rise of the Red Dragon* (2020).[18]

JP - So, there are things that are happening around the world that it's going to be tough when all these spaceships or UAPs start showing up on these different parts of conflicts and different types of volcanoes, and they're just going to show up and stop everything. And everybody's going to stop in their tracks, and that's how I think what's [going to] happen. That's why the Nordics got control of the ark because they also got control of the Lunar Ark as well. And there are different ships coming from the region where it's in between Ceres and Jupiter. So, we got different ships coming this way as well, and they're huge ships. There's going to be a lot of sightings of different types of ships, and people are going to have more of a green light when talking about this.

I know things that happen are probably negative, and it's sad, but there are a lot of distractions for this type of information to come out, and people need to think positively and think about love and share that love.... OK, so it's going to be exciting, Doc. I don't have as much information as I used to about the arks.

MS - Okay, well, is there anything you want to say to our Brazilian or Spanish friends listening to this? Final words?

JP - Sure, sure. I'm going to talk to Brazilians. Brazil, all the way. I think it's momentous, a lot of difficulties in the world. I think it's positive. It's a certain moment that I think it's all about searching. Brazil is very special to me. It's a lot of Brazilians. And in front of me, I think it's very important to have conversations. I think it's very positive. Now I'm going to talk to my Spanish-speaking people.

Hello, everyone. I'm from America.... I think it's very important that the situation that we're in in the world is very difficult because we know that there are some positive things that are going on. There are some negative things going on, as well as some positive things. It's very important for the people who are there, and it's very possible that they will be there. And it's very important to have information about Brazilians. It's very important to have a lot of information about things that are very important. All right, cool.
MS - Thank you, JP.

JP revealed that different extraterrestrial and Inner Earth civilizations, along with space arks, were going to be revealing themselves during tumultuous times for the entire planet. Their appearance is certain to shock and awe populations all over the world, and hopefully help countries involved in violent conflicts to pause and inspire them to seek peaceful resolutions. That, at least, is the hope, but as JP says, there are negative groups at work that want to confuse people and prolong international conflicts.

Chapter 3

Inside a Flying Saucer and Nordics take control of Space Arks

On November 4, 2023, JP encountered a Nordic-looking extraterrestrial who landed in a flying saucer-shaped craft near his home, similar to one JP photographed in 2018. The Nordic, whose name is Jafis, wore a distinctive baggy pants type of uniform, a style similar to what is worn by Indians or in the Middle East. Jafis told JP that the Atlantic Space Ark had relocated to the middle of the Atlantic Ocean, making it inaccessible to surface humanity as it was increasingly 'worshipped', and this was becoming a problem. Consequently, the Nordics were taking increased control over all the arks and restricting access to Earth militaries.

JP further said that he felt very peaceful inside the Nordic spacecraft and consequently accepted a drink with a blue substance inside it. Jafis told JP that the substance would remain inside his system for a long time as his bones would absorb it. The substance would allow the Nordics to better track and monitor JP. It made him feel very good and enlightened. Soon after, JP began getting urges to become innovative again and build healing technologies using crystals. Jafis became very emotional at one point and revealed to JP that a temporal war was causing much damage to different timelines. JP was told that he would be taken by the Nordics in a future mission to view some of the timelines that were being shown to the militaries of different nations, aka the Earth Alliance. The transcript of the interview follows, with grammatical corrections and my commentary.[19]

Interview Transcript with Commentary

Key: MS – Michael Salla; JP – Pseudonym for US Army Soldier

MS - I am with JP. It is Monday, November 6 [2023] and he had a visitation involving some Nordic extraterrestrials. So welcome JP to Exopolitics Today.

JP - How are you doing, Doc? I really appreciate this moment and time that we're living, and this moment and time that we're talking and bringing this beautiful information to everybody.

MS - So when did this visitation happen?

JP - This happened two days ago.

MS - So we're talking Saturday, November 4th [2023]. So yeah, can you describe what happened? Where were you?

JP - I was driving my car, and I saw a light following the back part of my car. I was like, man, what is this? What the heck is this? So, I pulled over, and it was already dark. It was night. It's around nine-ish. I pulled over, and I'm like, what the heck is this? So, I looked up, and I saw a beautiful ship. The ship that I saw in Orlando that I have a beautiful picture of it. The one that had the porthole. But this was that night. It was a beautiful, beautiful ship. One of my friends actually has a video of it. So we're still fighting to get that video to you to see if that video could be released to the public for them to see the ship….

The ship in Orlando JP is referring to here was sighted and photographed by JP on January 12, 2018, after he received a telepathic communication to go outside and look up into the sky. He says he could see a flying saucer with a large rectangle-shaped porthole through which he saw two Nordic extraterrestrials. The incident is described in *US Army Insider Missions 1*, and the photograph JP took is reproduced below.[20]

This is the third in a sequence of three photos taken by JP on Jan 12, 2018. The image on the left is the original and the upper right extract shows a close up of the flying saucer photographed by JP

Figure 7 JP photo of a flying saucer showing two figures behind a rectangular porthole

JP - The ship landed in the woods area. So, I parked my car and started walking towards the woods to see what it was. I felt a good vibe. That's why I walked to the woods. Similar to the same feeling I get when you visit the arks, you know, or when I went and touched the ship over there in Brazil. I felt that vibration, that same feeling. So, I start walking through the woods, and I see in the distance ... that beautiful ship. It was glowing a bluish-white color ... and then it dimmed down. It really dimmed down to a navy blue fluorescent color. It landed, and it was humming. So, I was like, do you hear me?

I did not hear any answer back. So, I kept walking towards the ship, and I walked towards the ship, and I see a Nordic there standing. He lifts up his hands, he puts his whole palm like that, and he does [a peaceful gesture] to me. I hear his voice, and he tells me, "I'm peaceful. You can come forward."

MS - ... So just for people who are listening to this and don't see you, you just raised your palm straight up. The Nordic just raised his palm straight up and said, I'm peaceful.

JP - With his fingers all together, just pointing straight up in the sign of peacefulness. Yeah, I'm peaceful. You can come forward. I did not hear a voice; I heard it in my mind. I'm sure they have the technology of telepathy. They're more spiritual, more enlightened, more technological than us. He had a uniform on. It was a military uniform. It wasn't a uniform that I had seen before, but he was dressed elegantly. No tights. It was more of loose pants, similar to what the Indians in India, wore back in the older days.

MS - Okay, yep, like the baggy pants.

JP - Yeah, Aladdin, but it was all one color. Looking like Aladdin. I said it in my mind.

What JP is describing her is very similar to what George Adamski described of the extraterrestrial Orthon, who claimed to be from Venus. Orthon wore a baggy set of pants similar to what JP described as common in India, and they were also worn by the Aladdin character. The following illustration of Orthon shows the kind of baggy pants worn by the Nordic that JP encountered.

JP - And he kept telling me all, "I'm peaceful, walk forward." So, I walked forward, and I looked at him straight in his eyes and he had a gaze, a healing gaze, beautiful. It's like when you look into his eyes, your whole body feels numb. And you're like, not hypnotized, but you're like, your eyes fixate on his eyes. And he says, "Come forward." So, I kept walking forward towards him. And then a platform came down from the ship. The ship was the size of half a school bus, but the height of a school bus. So, it was the size of a half school bus, but the height of the school bus.

MS - Just in terms of feet, we're talking ... 20 feet [6.1 meters] in diameter and about 14 feet high. Is that about right?

JP - Yeah, around there. More than 30 feet wide and round. If you measure it from each angle, it's about 30 feet. And it's a typical flying saucer like ship, but it was more smooth to the end of it. But it was a flying saucer-type ship. He tells me to come into the ship. And the platform comes down. He walks on top. I walk on top, and once you walk on this platform. And you stand at this platform. Your feet feel like magnets, so it [my feet] fixates on the platforms like you won't fall.

Figure 8. Illustration of Orthon, who met with George Adamski in 1952.

JP - And it goes up, and it locks. He said, "Sit!" I looked around, I saw a chair. It's like it's made with the ship. But you can clearly see ... the chair. So he told me to sit. I sat down and asked him a question. I said, "Hey, I know that you guys went to the arks. What's going on in the arks?" I felt like saying that to him. But I physically said it. I did not say it through my mind. And he told me that the ark was moving to the center of the Atlantic to position itself closer to a place where it was not worshipped, but a place where it used to be.

I thought, man, this is crazy. "So, is there a possibility that the military will revisit the arks again?" He's like, "The possibility is slim, but we're taking over all the arks. We're taking over all the arks." And now they're in charge of the Pacific Ark. So, little by little, they're taking over the arks, the Nordics. And they're moving more to the center of the ... Atlantic. And these arks are vibrating. He was telling me that the arks are vibrating. I don't know what that means. And then he said to me that he needs me. And that he was going to get in touch with somebody else from another mission.

And I said, "How did you find me? And how did you know I was around the area?" He's like, "We feel we know your vibration. We know we know where you're at at all times." So, some people have a certain vibration that ETs can pick up. And that's why sometimes they get followed all their life by these particular extraterrestrials because they monitor [humanity]. Their vibration is different from other people, and [so is] how they think and how they do things.

So, I'm sitting there, and I'm like, "I'm feeling happy. I'm feeling the ship." Once you sit in these ships, you feel like the ship is part of you. So you feel happy, you feel healthy. You feel the ship connected to you easily. And a lot of flashes, flashes through your eyes. Like when you're blinking, you flash, you see clouds, and then you open your eyes, and then you flash. You see other worlds, like beautiful landscapes where the ship has been.

And I told the Nordic, what's your name? And he told me his name was Jafis ... I'm like, "Okay, Jafis ... cool name. Awesome, is there a reason why you want me to go on the ship?" He said he wanted me to take and drink something that the ship had. So, he gave me a beautiful, it looked like a glass cup, but it was thicker, a thick glass cup and it had a blue substance inside of it.

And he said, "I need you to drink this." And I told him, "Hey, is this going to harm me? I feel really peaceful here." He said, "No, it's not going to harm you. This is going to stay in your system for a long time. And we're going to monitor you better with this liquid inside of you, and it ... absorbs into your bones when you drink it." So I drank it, and it tasted similar to a mixture of lemon and pineapple, and it left a taste, not a bad taste, but like a perfume, you know, and you're sprayed perfume in your mouth that bad sensation that you get in the back. It doesn't hurt or anything. It just doesn't taste ...

MS - An aftertaste.

JP - It doesn't taste nice. Yeah, [it was] an aftertaste that was not as pleasant, but he said it was normal to have that taste, and it goes away in a couple of hours. But I had that taste, like a dark chocolate aftertaste. And I felt good. I felt enlightened. I felt like I was getting more information. And I felt the urge, I think I told you, like two weeks ago or a week ago, that I felt like building a type of technology that I was building before. It's a type of healing technology. And I think after the interview, I'm going to ask your opinion about it. I can't talk about it right now, but it's a type of healing technology, but it's like a better version of it using crystals and other things that I think we can use, that we can find on Earth. So it's going to be quite interesting.

So yeah, he gave me that drink, and he got up, and he … put his hand down, and he got me on the platform. I went to the platform, and then he looked at me, and he started crying. And I looked at him, and I, in my mind, said, "Why are you crying?" He's like, "Because of what humanity has become, things can change dramatically in this timeline." And he said that to me, "So, there's something happening with … timelines that is happening now. That is really right now."

I think everybody should [have] … a positive and more lovely heart in this moment of time. Because when there's a lot of hatred, when there's a lot of things happening around the world, that's when timelines are getting messed up and certain things can come back. And certain things can develop that can damage some of the things that have been happening…. So, I asked him, "What do you mean about the timeline? What do you mean?" He's like,

"Soon you're gonna get picked up again, and we're gonna take you to the place that we're gonna show you about the timelines that we have shown to certain militaries around the world."

In a previous mission conducted on June 23, 2023, JP was taken to a facility where Nordics were helping the US military monitor four periods of time regarding the appearance of space arks: 1550, 1940, 2050, and 2090.[21] It appears that JP is here referring to a future mission with the Nordics, who will give him more information about future timelines.

> JP - So he went like this again; with his hand, he pointed to the platform. I looked at him, and I started tearing up because I felt what he was feeling. So I went down, he stayed up, and he looked down at me, and he just waved at me. I walked out and saw the ship; it closed up and went all the way up. It shined like a beautiful bluish-white light down into the woods. It turned everything into daytime. And I saw everything like daytime.
>
> It was beautiful. Then it just went; it just shot straight up, became so small, and disappeared. I went back, and there was a cop with the lights on behind my car. He was investigating why my car was parked there. I had really walked into the woods, and I left the car with the lights and everything on. It was in park [mode], so the cop stopped me. He says, "Oh, I thought nobody was in the car. I checked the gas, and it was good." The cop was telling me this. "Are you okay? What were you doing in there? Did you see that?" I'm like, "You saw what?" He's like, "Did you just see that in the sky?" I'm like, "Yeah, I saw that." He's like, "Oh man, that shit was crazy."

I wanted to ask his name to get his testimonial. He's like, no, no, no. I'm not going to get into that. Just drive safe. Next time, put your blinkers on. If you leave your car like that, I don't know if you're taking a piss or shitting over there, but next time just, and I asked him again, "Did you see what I saw?' He's like, "Hey man, I don't know, but stay safe." I'm like, "all right, cool." So, the cop went to the side, and he left. I forgot to ask for his badge number. It would have been good if I had his badge number. I got in the car, and I started heading back home. And it was a beautiful experience with the Nordic. I haven't had an experience like that since Orlando.

MS - So, the craft in Orlando that you took a photo of had that big panel, and it had two figures in it. That's the same craft?

JP - Yeah.

MS - Okay. So we got a picture of that from Orlando. So that's good. And you said there might be a video of this?

JP - There might be a video. There might be a video of this ship taking off.

MS - Someone else took a video, and we might be able to get it.

JP - Yeah, that'll be awesome if I can get that.

MS - Yeah. Okay. Well, too bad the cop wasn't willing to talk about what he saw

Nordics Take Control of the Space Arks

In the following part of the interview, JP reveals what Jaffis told him about Nordics taking over the space arks found around the world's oceans that are being moved.

> MS - So the arks are moving in the ocean…. We know there's one in the Pacific. We know there's one in the Atlantic. But the other arks you've described, Caldas Neuvas, Ukraine, and Russia, are all under land. So they're, I assume, not moving. But the Nordic said that they took control and they're moving the arks. So, did they take control of all the arks, even the ones in Ukraine, Caldas Nuevas, and White Sands in New Mexico?
>
> JP - They said that they're going to take control of all the arks. But now they're really in control, the ones in the oceans. I don't know about the ones in Ukraine, Russia, or Brazil.
>
> MS - And did he explain to you why they needed you?
>
> JP - Why did he need me?
>
> MS - Yeah, why did they need you? If they're taking control of all the arks, why did they need you? Did he explain that?
>
> JP - No, he did not explain why he needed me. It did not cross my mind to ask him. But when I was visiting the arks, I remember activating certain things in the ark that a lot of people were not able to activate.

JP - And because of my vibration, because of my, I guess, way of thinking, I think they know that my past is from somewhere else. I know some things that will be important for them too.

MS - You're connected to the arks in some way. Do you remember when you went back to the Atlantic ark for the fourth time, and you took the crystal, and you said you guys got out and you were gone for like two days, but you thought you were only gone for two hours? And then you had flashbacks. You and other people had flashbacks of ETs in the ark. Do you have any more information about that? Have you remembered more about who those ETs were? I just want to know if they were ETs, if these Nordics were taking control of the arks, or if they were like the crew.

JP - Well, I've been having flashbacks here and there about different types of ETs interacting with the ark. These arks came from somewhere else, you know, and different types of ETs and civilizations also have a connection with these arks. It wasn't just one race that built these arks. So, it's a mixture of races that built these arks. And I think there's, it's not just on Earth that there's arks. I think there are arks on other planets as well.

JP - You know, they're like beacons. I'd say beacons for other planets that have these arks that are buried there as well. So, each planet has its arks, just in case something happens to their planet.

Elena Danaan has also talked extensively about the space arks and said that approximately 22 out of the 24 Seeder races used arks to seed ancient civilizations on Earth. The arks would provide a technological foundation for the new civilization and could also be used as an evacuation vehicle in any planetary contingency that might emerge. In her book, *The Seeders*, Danaan received a communication from Thor Han Eredyon about the purpose and history of space arks in our solar system:

> A long time ago, the Intergalactic Confederation had several colonies in this star system. On Naara (Venus), Terra, Luna its moon, Tyr (Mars), and on the fifth planet. Great wars occurred with the Anunnaki, the Ciakahrr, and local colonies, whose civilians had to evacuate to Terra or to outside this star system. Before leaving, they gathered the essentials of their knowledge in arks they buried deep under the surface of the planets I mentioned. These arks preserved the essential information necessary to rebuild the glory of these colonies, if one day this was to happen.
>
> There are diverse origins for these arks, either that they were carrying refugees from the Sol wars, mainly from Mars and from Maldek, the former fifth planet, or either from colonies of diverse origins. Refugees from the Man *(Lyrans)* wars also arrived in these arks and built their ground colonies around them. This is the reason why you will rarely find an isolated ark; entire cities were built around them, and in some locations as well, you can find what the Terrans call: "Halls of Records." Pyramid generators are also to be found in the vicinity of these Arks. Many arks are interfaced with each other by portals. When the fleet from the Intergalactic Confederation approached this star system, the arks activated.[22]

My interview with JP next discussed the identity of the Nordics inside the Atlantic Space Ark, which he had encountered in previous missions to the ark. He described his fourth mission to the Atlantic Space Ark that involved the extraction of a crystal jewel in *US Army Insider Missions 2*. His fifth mission was described earlier in Chapter 1.

> MS - Okay, but as far as those ETs that were inside the ark, when you went into the Atlantic Space Ark to get the jewel.
>
> JP - Are you talking about the Nordics that stayed behind?
>
> MS - Now, that was the fifth mission. In the fourth mission, you said that you guys were in there for two days but couldn't remember being in there for that long, and then later, you had flashbacks of some ETs in there. And I just wanted to know if you remember anything more about who those ETs were?
>
> JP - No, I don't remember who those ETs were. I know that certain sections of the arks are run by different DNA, and different ETs work together. But no, I haven't got any information about the flashbacks about the other ETs. Being honest here, you know.
>
> MS - And you said that with your fifth visit to the Atlantic Ark, there were the four Nordics that stayed in the ark after you and the other guys, the four of you soldiers got out, that the four Nordics that were with you stayed behind. Were they dressed the same way as this Nordic in the craft that just picked you up?
>
> JP - No, they were dressed in a military type of uniform. But it is the same type of Nordics. Yeah. So

you have the Nordics that are connected with the military, and then you have the civilian types of Nordics that connect with the civilian sector. The ones that are connected to the civilian sector, some of them wear whole body tights or they look like they'll dress up like the baggy, Aladdin type of pants. And their clothes are really loose. And this Nordic that visited me, he was like six, probably like 6' 6", 6' 5",

MS - And again, blue eyes, long white hair. ...

JP - His eyes were whitish blue. Really nice eyes. Yeah. And inside the craft he was the only one inside that craft. He was the only one that I saw. Okay. I'm sure there was probably one more somewhere, but it did not come out. I really sense two beings inside the craft.

MS – Okay, so the craft was about 30 feet in diameter, roughly the size of a school bus, 12 or 14 feet high. But when you're inside, did it look any bigger than how it looked from the outside?

JP - It's weird, but yeah, it looks bigger. It does look bigger, and time does run differently there. Time is different in these ships, and they have little compartments. It's separated, not walls, but they're ramped up, and they separate into triangle walls. So, it could be a size 12 by 12 feet, and then 12 by 12, 12 by 12, 12 by 12, but it's separated. And you get through these different rooms through the middle. Like you can curve to another room, you can curve to the other room, you can curve to the other room. And it's really lit up. So, everything's like a whitish-bluish color.

JP's observation about the discrepancy between the external size of the flying saucer and its much larger interior has been noticed by previous experiencers and whistleblowers.[23] It appears that the interior of the craft is operating under a different time-space continuum that affects both the flow of time and the interior size of the ship. This discrepancy is very likely due to the extremely high electrostatic charges used by ET spacecraft that impact space and time.

> MS - So you were just in one room, but there were other rooms.
>
> JP - Yeah….
>
> MS - There could have been other Nordics in other rooms. You said you sensed two of them.
>
> JP - Yeah, I sensed the one that was talking with me, who was there in the same room with me that picked me up. And then there was another one.
>
> MS - Now it seems odd that he would give you a drink and then pretty much let you out of the craft. I mean, that seems odd. Normally drinks are to prepare you for something. So why did he do that? Why did he give you a drink and then say, "Okay, you're on your way?" Why was that? Did he give you any idea?
>
> JP - No, he did not give me any idea. He just told me to drink that particular drink. And then, he told me that this drink would help them track me better.
>
> MS - I see. Okay. All right, well, that sounds like …

JP - I guess it connects me more with them in [terms of] communication. I think it's more of a way for me to have a feeling, a way to connect with them, being in a far-distant location.

MS - Okay. Well, it'd be good to know if in the next mission with those ETs, with that Nordic, if he's the same one that picks you up, if you find out more about what planet he comes from, if he's part of an organization, it would be an interesting question to ask.

JP - Oh, yeah. I agree. I think a lot of people are going through these types of experiences.

MS - Okay. Well, anything finally you want to say to the audience, maybe to our Portuguese and Spanish friends?

JP - ... I'm going to talk to the Spanish-speaking people. You live in the farmlands in Mexico, South America, Peru, Guatemala, Colombia, Argentina, and Chile. Those people living in the farmlands where there is no light pollution. You see stars, right? Good. But you, at this time, in this month, have received visitations from different lights. Don't be afraid. Receive them as if they were happy because they are a kind of blessing. In Brazil, South America, and Spanish-speaking Latin America, everybody that's living in the farmlands, look up, they're going to start receiving visitations of these Nordic types of ETs. People are going to start coming out and talking about these visitations, also in Europe or all over Asia. They're going to start showing up, Doc. And they're going to start visiting certain people that have this type of vibration.

MS - That's awesome. Well, thank you, JP, for the update on your visit.

JP - No problem. I just want to say to everybody, just think positively. Bring love, a lot of love, share a lot of love. Do kind things to others. Don't do bad things to people, or to yourself, or to your body. Be nice. Bring positivity to the world and share the love. Only with a simple hug, with a simple smile, you could change somebody's life. I think humanity means that right now. With the things that we're going through.

MS - Amen to that. Thank you, JP.

JP - No problem, Doc.

The news of the Nordics taking control of the Atlantic Space Ark was something that generated controversy. My colleague, Alex Collier, publicly expressed his reservations about this development soon after JP's November 4, 2023, update was released.[24] The core issues were should the Nordics be regarded as truly friendly if they took control away from humanity of such vital technology, and was Jafis part of an unfriendly Nordic faction manipulating JP? On the contrary, was the Nordic takeover of the ark and moving it away from the Bermuda Triangle area a positive development as it would lessen US influence over the ark and make it more of an international concern? Was JP being helped by Jafis and other Nordics in dealing with growing health issues arising from his covert missions to the Atlantic Space Ark and elsewhere? These are critical questions which are addressed in Chapter 5.

Chapter 4

6th Mission to Atlantic Space Ark

On November 28, JP completed his sixth mission to the Atlantic Space Ark, which was moved from the Bermuda Triangle further towards the mid-Atlantic into deeper waters in order to create greater international cooperation. He says he was taken again by a TR-3B craft to the donut-shaped naval ship that has been positioned over the ark since at least 2015. An elevator tube from the donut-shaped ship was still in the process of being re-attached to the space ark, so JP says that a submarine/spacecraft with large hemisphere-shaped radomes attached to it transported him and his team down into the Atlantic ark.

JP said that the Nordics, along with Inner Earth beings, had taken over control of the ark and were now activating it far more quickly than when it was under the control of the US Navy. He saw 18 Nordics & Inner Earthers who were dressed differently walking around the ark but believes there were probably more in other areas they did not visit. The Nordics were particularly interested in what JP knew about what the Aztec Indians had activated during their earlier visits to the ark.[25]

Upon the mission's completion, JP told the officer in command that he didn't want to do future missions due to the personal toll they were taking on his body. The commanding officer was extremely angry, and upon returning home, JP was followed and attacked with some kind of frequency weapon by individuals in a black van. The next day, he ended up in the hospital. This is the third time that JP has ended up in the hospital after doing missions, which has made him reluctant to do future missions. The transcript of the interview follows, with grammatical corrections and my commentary.[26]

Interview Transcript with Commentary

Key: MS – Michael Salla; JP – Pseudonym for US Army Soldier

> MS - We are back with JP and he has had another mission. It is December 2nd [2023], so welcome JP to Exopolitics Today.
>
> JP - Hey, how are you doing, Doc? I'm glad to be here. I'm glad to bring this information to you. I got green-lighted really fast for this mission to talk about it with you. Quite interesting.
>
> MS - That's great. I'm very happy to hear that. So yeah, what are you telling us about what you experienced?
>
> JP - Yeah, sure. So we actually drove to an open field. It was a total of five of us, five soldiers. We were dressed in civilian attire, and we were there in the middle of the field, and we saw a TR-3B coming in. So, this TR-3B was the one that picked us up and drop us off. Sometimes, the pilot doesn't come out, or the pilot doesn't say hi or anything. We just know they're there like a transport to pick us up and drop us off. The fast-paced type, like really fast.

JP is here referring to a triangle-shaped craft, which is a smaller version of the TR-3B, which has edges that are 600 ft (183 meters) in length. JP described the flying triangles he had been on as being more the size of a school bus, with edges approximately 40 feet (13 meters). JP took photos of many of the flying triangles that are reproduced In *US Army Insider Missions 1*. The following diagram features a flying triangle photographed by JP on February 18, 2019, which he calls a TR-3B, which is similar to the much smaller craft he was taken on during this particular mission.

> JP - So we were waiting in the field, and we saw this TR-3B coming down. It hovered. There was a guy on TV a little bit more down doing security, and they saw it, and they flashed a light, like a light beam, to the TR-3B, and it knew the location where to hover on top. There's a type of code that is similar for certain types of ships, like the TR-3B or flying saucer types of ships or other ships that the military uses to maneuver and pick up people and transport them.
>
> They flashed the light to the TR-3B, and the TR-3B started hovering over us. So, it went down. It was six feet above the ground, and we just entered through the back part of the TR-3B. A little ramp came out, a grayish-black ramp that came down, and we got on top, and it took us back into TR-3B. So, we got in there, the TR3B, and we started going to the Atlantic where the Atlantic ark is. This is quite interesting because we saw the ship; the donut-shaped ship was already at the Atlantic Ark's location.
>
> JP - So right now they're in the preparation of bringing down the tube, the elevator back to the Atlantic ark. But when we got there, it wasn't prepared yet. So, there was a huge, huge ass submarine there. The submarine had a radar bubble in the front, but it was huge. This submarine was huge. It was waiting for it by the donut ship. There were other ships there as well, international ships. I can't tell you the other international nations that were there, but there were four nations that play big ball around the world that they were there.

4th image in sequence of four photos taken by JP on Feb 19, 2019. Original on left, enhanced version on right.

Figure 9. Photo of Flying Triangle

MS - JP, can you explain what you mean by a big radar bubble around the submarine? I haven't heard of that before.

JP - So it has like a metal black bubble looking. How can I explain this? Okay, it looks like a ball, but the size, the ball is the size of an ice cream truck, the size of an ice cream truck, like this is round and it's huge. And this is a type of radar that they use when they go underwater or this particular type of submarine/ship goes into space. It's a submarine, but it also can go to space.

MS - I see. And this kind of spherical radar is at the very front of the ship, like at the ...?

JP – It's at the very front, and it also has one at the bottom. But the one at the bottom is smaller. It creates a vibration around the ship, and it can even levitate out of the water. I think we use these types of submarine-looking [craft] as spacecraft to also go into space. And I saw one of them. And that's the one that we entered. So we entered into this big-ass submarine, and you know, I had never been in one of these ships before.

There was one incident over there in Tampa. I remember getting on a ship that was like a flat platform type of ship. It looks similar to that type of ship, but it's more ET-like looking. It wasn't more military. You could tell it was more ET-looking. The way I compare ... ET and regular military looking [ships, is that] the ET ships look like they're fabricated all [at once] looking like one piece. On the military-looking ship, you can see parts moving around. You can see different parts that are connected after the ship is built, and all that, but the ET ships are built like one piece of ship.

JP is here referring to the different construction methods used by extraterrestrials for their spacecraft when compared to those built for the US military by major aerospace corporations. While the former is grown in space as an organic unit that possesses a form of "plasma consciousness" connected to the crew, the US corporate-built spacecraft show clear signs of distinct metallic parts welded together or joined by rivets as exemplified in modern rockets or naval ships. According to David Adair, he encountered the engine of a large extraterrestrial spacecraft in 1971, hidden in a subterranean vault at Area 51. He says that he encountered the ancient consciousness of the engine, which he called Pitholem, that decided to use his body as a lifeboat until it could be rescued by an incoming fleet of ancient motherships that appear to be similar to the Intergalactic Confederation (aka Seeders) described by Elena Danaan.[27] In the final chapter, Danaan explains how space arks have an organic 'plasma' consciousness that is vital for powering the craft and that they can also be downloaded into crystals for storage purposes when necessary.

> JP - So this ship looked more ET-ish, but you can see military people working it out and flying it, and it goes underwater, and it can go really deep underwater, Doc, this particular submarine-looking ship.
>
> MS - Okay, so even though it looks like an ET ship in terms of its appearance, the crew is a human crew.
>
> JP - Well, I'm sure that this ship looks like a cigar-looking ET ship when it's flying. It has the characteristics of that type of ET-looking cigar ship, but it was like a submarine when we got in. It was underwater. It was on the water. So, the way we got in it, the ship totally came out. So that's why I could see the size of the ship. You can clearly see it coming out. And then I heard one of the ships from the international ships honking the horn, like, Burrr! It was really loud when the ship came out from the water. So, this ship has a type of anti-gravity [propulsion] because of the weight. The amount of the ship that came out. It would have been too heavy for the water not to hold it down.

MS - How long was it?

JP - It was around 600 feet (183 meters) around there. It was huge, Doc.

MS - Okay, so like two football fields. It's kind of like an aircraft carrier, like a small aircraft carrier [in length].

JP - Yeah, similar to an aircraft carrier. Yeah, it was a huge, huge ship. It was big. It was almost the size of the donut-shaped ship that we had floating there. It was a huge-looking ship. And it floated out, and we got in it. It has three floors of compartments. It's quite interesting. So, we went down, and the ship started vibrating. You could feel the ship vibrating. You can feel everything vibrating going down towards the ark.

I felt kind of happy that we were heading back. But the Nordics are in charge of the ark now. So the ship had little port holes that you could see outside, but this type of glass was really thick, and you could still see lighting coming in or coming out. And then we were getting close to the ark. You could see a light that the ark never had. So, the ark was kind of activated in a different way than we had ever seen before.

MS - And the Ark's position is that it's in a different part of the Atlantic Ocean. Before, you said it was being moved into deeper waters. So, was it, you know, kind of like deeper waters further away from the US coast?

> JP - It is farther away from the US coast. So right now, more international ships can go around it. We saw more international ships around the donut-shape ship, and that was quite interesting. Yes, it's farther towards the middle of the Atlantic and it took maybe a month or three weeks to get to that location. It got to that location through a vibration that the ship was going through. They [the Nordics] put it into a vibration state that made it move.

In JP's previous mission to the Atlantic Space Ark on September 11, 2023, just as Hurricane Lee began impacting the Bermuda Triangle area (see Chapter 1), he described the ark as beginning to move at a couple of feet per minute and the Nordics staying on it to activate dormant systems. To arrive at its new resting place in the mid-Atlantic in three weeks, the ark must have accelerated.

> JP - But now the ship seems more activated. There's more lighting on the ship, surrounding the ship. So the submarine-looking cigar-looking ship is actually run by the [Nordic] Navy. So that's quite interesting. And we landed on top of the ark and then locked. And we can hear like a popping noise on the [submarine] ship. The whole ship moved, and we had to take a ladder down into the ark to a location that we had not been to [before] that had a lot of vegetation.
>
> We saw a lot of light, bluish-white light that was all over the place in different corners of the [interior] location of the ark that we were in. And we saw at least six Nordics walking around with what looked like cell phones, but are actually sensor types of machines that they have, that they're walking around with. Just constantly walking around and

working on information about the ship. So, we landed there, and we got into this part of the Ark that was there, and we started going to familiar spots that we went to. And when we saw these places, they were more activated. They had more vegetation and more light. So, whatever the Nordics were doing, they're doing a good job at it— activating the ark more than it was activated before when we had it. So I guess they have more knowledge on the activation of these [ark] ships.

MS - And when you say "we", you're talking about the five of you that were taken there to the donut-shaped ship on the submarine. Was it just the five of you that went down, or were there others who joined you?

JP - There were other people that went down, but they couldn't go farther than when we went down with them. So, when we go down, we usually go down; we go more into the ship, and it activates, but the ship looks like it's activated. It's really activated right now. The ark has a type of vegetation that I haven't seen before. It looks like grass, but it grows in a curly way. Like a curly fry way. It's not straight, like regular grass. It grows round, like curly fries. So that was quite interesting. And that's all over the wall. You can see that type of vegetation over the wall. I think it's an extinct type of grass that the Earth had that doesn't grow anymore, but still grows in the [ark] ship. Quite interesting.

MS - Was there any kind of analysis of the vegetation growing on the walls? I mean, from earlier missions, did anyone, a biologist or a botanist, examine the grass or the plants growing on the walls? And did they say what it was?

JP - So these plants were not there when we were doing the ark missions. These plants are now there, and the Nordics took over. So it's more lighting and more vegetation right now. So no, these plants, we never saw them [before]. I guess they're like chia plants that they grow when there's a certain lighting or a certain activation on the ship... And they give out beautiful oxygen to the inhabitants of the ship. So it's quite interesting. Certain types of plants that are growing in the ark are already extinct, but they're growing now in the ark, and it's really, really, really nice. Different. We haven't seen the ark this activated. And one of the guys was talking to a Nordic, and the Nordic was talking in regular English like he did not have an accent. He sounded like an English [accent] from New York, like he had that New York English accent.

MS - Okay. And was it just the six Nordics that you saw on the Ark, or were there more as you walked through the Ark?

JP - Oh, there ... were more Nordics there than we saw. We saw more than 16, maybe 18, Nordics walking around. And I'm probably sure that there were more in different parts of the ark and all that. They were talking to us, and then there was a sharing. We were just collecting data and exchanging information about what we have with the ark before and what we do [now]. So we're just having a type of meeting, but walking around as we were talking, and the device that they have writes everything down for them in a certain way that they can have it [recorded] as information. So, we're just sharing ideas of past missions.

They're really interested in how the Aztec Indians came in. And they were really asking questions about that. And what did they activate? They [the Aztec Indians] seem to have activated something really important in the ship [ark], and nobody knows how they did it, but they did it. So they were asking for information about that. They were asking if we could find them again to bring them down. And for them to talk with the Nordics about it. So we're in that process now to see if we can find the [Aztec] Indians that activated the ship more. So that's another thing happening right now.

In his first mission to the Atlantic Space Ark, JP explained that the Aztec Indians were ecstatic to be in the ark as they believed it was the fulfillment of ancient prophesy. He described them as singing and chanting "A Kuria Matte", and how all this allowed them to access deeper areas of the ark, which the Chinese and American soldiers could not access. This led to the US and Chinese leaving the Aztecs behind, and 'rescuing' them in a second mission that was jointly conducted with Russian soldiers.[28]

> JP - And the Nordic that had that New York accent was talking to another soldier. He was telling [him] that they found a vein system around the ark. It's like a vein that has a substance that is blue, a bluish liquid substance. And I remember having another mission before about me drinking a blue type of liquid substance. And now they're finding out that the ark has a vein system that goes all around the ark, and it gives it energy. All around the parts of the ark, there's vegetation. There's light. So it's quite interesting.

MS - Okay, so you're talking about your previous mission where you were given a blue drink by , and that made you feel better and more active. And you're saying something like that is happening with the Ark as well?

JP - Well, this substance was always in the ark, and these veins are connected to the ark. So, remember I told you that the ark was like an organism that's alive, that's always living, that it could feel you, and it could feel who comes in the ark and who doesn't come in the ark and all that. So, this substance is all around the ark. There's a type of liquid, but it's in tubes and in types of silicon-looking veins that are integrated inside the ship. The Nordic found out that the ship had the veins they spread out. So, where the [crystal] jewel is, starting from the jewel department, these veins spread out, and the jewel activates the liquid of the ark. And I think it keeps the ark healthy. I don't know how to say this, but yeah, I like for it not to deteriorate and not to break down. So. Yeah.

MS - And these veins, I mean, this is something you didn't see in your previous missions on the arks. It's just that now you're seeing them, and you think it's because of increased activation due to the Nordics?

JP - Well, these veins are integrated into the walls of the ark.

MS - But you never saw them before. This is the first time you've seen them?

JP - I have not seen these veins before. We did not see these veins before. Something has happened, and it must be the activation the Nordics have been doing to make those more prominent. Yes.

The veins distributing a bluish-white liquid to and from the crystal jewel that JP witnessed help confirm claims that the space arks are organic entities grown in space. The crystal jewel that is eventually inserted into the ark contains a plasma-like consciousness, which has been specially purposed to align with and recognize the crew's DNA and consciousness. This would explain why some individuals are able to access and activate the ark while others can't.

JP - So the Nordics are doing a good job activating the ark. And I'm sure they're doing it with all the other arks around the world and in our solar system. So it's quite interesting. It's not just Nordics. There are other ET races that are helping out as well. There's Inner Earth as well. They're helping out.

MS - Did you see Inner Earth beings down there? Did they introduce themselves? How do you know they were Inner Earth and not Nordics?

JP - By the way, the Nordics were talking about it... We saw more Nordics there, but those Nordics were dressed differently. There were Nordics with different uniforms in the Ark, but there were still Nordic [in appearance]. I don't know if it's different branches of what they work on or different types of locations; they are from different parts of our solar system. But I know that Inner Earth is involved with the activation of the arks by the missions that I went into the caves over here [in Florida]. The mission when I went to the Ant People..

There are certain things that look the same as it looks inside the arks. Like the vegetation growing on the walls, when we went on the mission to the Ant People, we saw similar plants that looked really similar to how the Ant People did it, and how it looked inside the ark was similar. So yeah, that's how I know there's a connection. There has to be a connection. Does that make sense, or my English is still messed up? I'm trying to, like, really put it into words and to try to make sense of it all, you know, in English. I'm sorry.

MS - That's okay. You just describe it as best you can; that's fine. So, you saw the veins, so what else did you see as you walked further into the ark?

JP - The gravitational ball of water where the fishes are, they're activated again. So, they're back [inside the liquid bubble] and not messed up, no more. The gravity is good. And I'm sure there were at least seven Nordics in that part where teleportation happens that can be teleported from one ship to another ship.

JP - So that's really interesting. I think they're working on that teleportation device that is on the ark; that's really interesting, as the military also wants information about that, about how it's working, and all that.

MS - ... In the past, you said that there were two spheres. One was with the fish, the fish were in it, and the other one was the teleportation device where you saw ... the 15 guys, soldiers?

> JP - They were like holding [on together] like into a chain [formation], and we were trying to get them out because they were starting to drown and all that.
>
> MS - That's right.

JP was here referring to a teleportation system that linked all the arks. In his second Space Ark mission, he described how he was part of a team of 20 scientists and soldiers that rescued another group of 15 soldiers from a multinational mission underway on an ark found on the Moon. The soldiers had stepped into a sphere-shaped portal on the Moon Ark at the same time as the Atlantic Ark mission was underway, but got trapped inside the liquid-like substance inside the spherical portal. JP learned that all the arks were linked by portals, which Nordics in charge of the Atlantic Space Ark could now use to travel between arks all over our planet and the solar system.

> JP - So, yeah, that particular spot we went to, and you could feel a beautiful vibration, and you could see the bluish ball of water right there in the middle. And you could tell that they were studying and looking at it. I think the way it works, you know, everything is done with vibration. Everything is done with the study of light...
>
> MS - Frequencies?
>
> JP - Yeah, Frequencies, the study of light, the study of the quantum world, because, you know, in order to teleport a particular person or a particular being, there are also living things on your body as well that have a certain DNA as well. There's a lot that goes with it when you think about teleportation. The way they're doing the study is [to find out] how the

teleportation device brings all the molecules back together, including the living things that live on your body. How many types of organisms live in your mouth or inside your gut? You know, if you do the study, there are different organisms that live in your body that have a certain type of DNA that is different from your DNA. So, when teleportation happens, it has to register all types of living things on your body and your body at the same time. So, it eliminates everything that is negative. So, it's a big, big science [that goes] into teleportation.

We're talking about heading back to the donut-shaped ship and the way teleportation works because it has to register every type of DNA and every living thing that is in your body as well. Not only a human being but also the organisms that live inside of you and help you with your body. You know what I'm trying to say? I don't know if it makes sense what I'm trying to say right now, Doctor.

MS - Yeah, maybe you need to rephrase that.

JP - So in order for teleportation to happen, right ... your body has different organisms that live in your body, right, microscopic organisms, whatever machine that you're going through has to register the DNA of those organisms as well as registering your body, your own DNA. Did that sound right or...?

MS - Okay, yeah, I can understand that.

JP - Okay. So that's what they were talking about with this device in the ark. How does that happen? Because I think that's also [the same] with time travel, you know, because if you do time travel as well, you're teleporting what, maybe, diseases or anything that your body has into that certain timeframe. It has to register everything that is on your body. It's not just your human DNA. It's more than that. So that's why it's so complicated. And it's hundreds of thousands of years more advanced than what we have as humans. So, it's really, really ancient. They were talking about that in the ark, and that's what the Nordics were studying. How does this device register not just the DNA of the person that's going but also the DNA of the other organisms that live on the person? I think people will understand Doc about that.

MS – So, it's an advanced teleportation system or portal system that the Nordics ...

JP - It has to be. Yeah, it must have been. So yeah, depending on how it works. And it did work because we registered [confirmed] that it did work. It does have the technology of teleporting everything on your body and in yourself. So that's quite interesting what they were talking about. And, the ship has other technology that they're [also] trying to figure out how it works. They have a regeneration technology that you can go inside. It's like a healing type of cocoon. I remember telling you that in one of the missions that we went on, we saw sarcophagi.... A couple of us in one of the parts of the ship saw dozens of these in different sizes of boxes, and these boxes are a type of regeneration [chamber]. It generates healing as well. And... it can also generate long stasis sleeping in it. So, they were also studying that [technology].

MS - Did you see, or did you know, or did you hear anything about anyone or the beings inside of the sarcophagi?

JP - No, we did not hear anything about that. We did not ask. We're just letting the Nordic do the investigation themselves, but they're sharing the information with the military. And I think they also got a connection with other international characters that were also there. So everybody on top of the ocean, the Atlantic Ocean, was waiting for information on what's happening with the Ark.

MS - Now, is that one of the reasons why the ark moved more towards the middle of the Atlantic, rather than being closer to the US, that it would be more of an international initiative to explore it?

JP - Yes, that's one of the reasons for that. I think it was the Nordics that brought that idea up, not to have conflict with other nations and all that.

MS - And can I guess who those nations are? I assume it's countries like China, Russia, France, and Britain.

JP - Yeah, totally. Yeah.

MS - Okay, so that's correct.

JP - I can't say if that's correct or not, but yes.

MS - Okay. All right. Well, that gives us an idea.

JP - Yeah, it gives people an idea of what's happening internationally and why it's happening.

MS - So it sounds as though the Nordics and the intelligence on the ark want to use the arks as a means of promoting international cooperation.

JP - Yes, trying to get everybody together and making sure that the right technology is given to us in order for us to prosper and to get better around the world, right... The Nordics are like mediators. So they like to put everything in order and make sure that everything is done right, and by the book, you know. It's not crazy like the Reptilians used to do ... I guess it's [more] organized, the way they [Nordics] do things.

So, by bringing the ship more into the middle, it's more safe. It's safer for other international nations to get close to the ark and also run their investigation. Because remember, they have arks as well. So, they want to share their information with this particular ark. This is the most activated ark on Earth right now. So that's quite interesting. Whatever they do with this ark, they're going to do with other arks. So that's quite interesting.

MS - Right, I remember you saying that the Atlantic Ark is the biggest of all the arks on Earth at the moment.

JP - And the second [largest] one is the one in the Pacific.

MS - I see; well, you haven't talked much about the Pacific one, but yeah, tell us, what else happened with the Atlantic Ark before we talk about the Pacific [Ark].

JP - It's more activated. That's one thing I want to tell the public that it's more activated. When this ark is activated, it creates a type of energy, and it's closer. It's closer to the middle of the Atlantic. So that's quite interesting. Okay, so it's no longer in the Bermuda Triangle. It's now closer to the middle of the Atlantic, and the purpose is so that would create greater international cooperation. Yes, we gather information. We said our farewells; we went back. We went back on the submarine-looking ship. We went up, and we went back to the donut-shaped ship. This huge ass freaking donut-ship. Can't believe that there are no pictures or there's nothing about it. I think soon the information is gonna come out before I think the ark comes out. This type of ship is huge. And the way it stays stationary with a big ass anchor that it has is quite interesting.

In *US Army Insider 1*, JP shared an image that he believed was very close to the donut-shaped ship and asserted that it was a form of soft disclosure as it was released as part of the Star Wars franchise of television series to openly reveal a highly classified US naval platform with minor modifications such as rounded edges. While it was used as a prison ship in the Star Wars series Andor, JP says that it is used for deep sea operations such as accessing a submerged space ark or city.[29]

MS - Now you say an anchor ... [in the] middle of the Atlantic. What depth are we talking about here? I mean, can ships have anchors that go to the bottom of the Atlantic Ocean in the middle? I mean, I don't know what [depth] that is. Is that two miles or something?

JP - I don't know how they do it, Doc. It's not a type of anchor that connects to the ship, but it has an anchor. I know that the ship is stationary. How the hell is it stationary? I do not know, but I know they talked about an anchor. A type of anchor that is connected to the ship, but it's connected like, I don't know. I don't know how to explain it.

Figure 10. Narkina 5 Prison Complex appeared in Disney Channel's Andor - Star Wars TV series

MS - When you say anchor connected to the ship, are we talking about the elevator or the tunnel that goes from the donut-shaped ship all the way down to the ark? You said that they were still setting that up.

JP - Yeah, they're still setting that up, but I don't want people to get confused. They do have a separate type of mechanism or anchor that helps us to stay stationary. But the elevator is something different ... [it] connects. Yeah, I think they do connect it with the anchor, but the anchor, I think it goes first. And then they connect the, it's like a tube, it's weird. It's not only that, but it's advanced [technology], which is really good. I hope people don't get confused with that, but yes, we do have technology that is way more advanced than what we have in the public. And yes, they do have a [advanced] type of anchor; the elevator connects to the anchor, and then it connects to the ark. And that's how we do our missions.

MS - Alright, so did you do a debriefing after you finished the mission? I mean, I don't think you actually explained what the mission was. So, what was the mission, and was there a debriefing?

JP - The mission was to gain information on what the Nordics were doing to bring that information back to the surface. And I was telling them, I was like, "Hey, I don't think you guys need me anymore. You know, I don't think you guys should use me anymore." And they kind of got mad because of that. When I said, "I don't want to help anymore." You know, I don't feel like doing these missions anymore. Then, the TR-3B picked us up, and they dropped us off. We got picked up by the van, and we came back. And then it was time to go home.

I was actually followed back, and I want the public to know that—to be careful. If you are a whistleblower, be careful who you [talk that's] not a whistleblower. If you have information that the public needs to know about these types of missions and these types of experiences, be careful who you talk to and who you tell this information to. Once I told a certain person on the donut ship that I didn't

> want to do these missions anymore, he really got angry at me. And I was like, you know, "I just don't want to be forced to do it no more. I don't feel like doing it anymore." You know, it's affecting me in a way that the only person I [can] say this to is you. And the reason I'm telling you is because I got the green light to tell you. So when I was heading home, I was being followed by a black van. And I got attacked. There's a technology that's out there that people use to get rid of people. And I guess they tried it on me, and the next day, I ended up in a hospital. So yeah, just, um, be careful.

In 1975, during a public hearing conducted by a US Senate Committee investigating the CIA's MK-Ultra program, the Committee's Chair, Senator Frank Church, held aloft a "heart attack gun" that could induce death in minutes without leaving a trace.[30] This was only one example of advanced technologies that could induce a range of health crises in targeted individuals without leaving any evidence.

> MS - So you think that was a retaliation, or was it a warning of some kind?

> JP - I know they're going to get mad if they hear this. I think it was like a warning. For me to keep going or, you know, not to stop doing what they're telling me to do. But, you know, I'm just going to keep my distance. I'm going to keep my distance and do what I need to do until I don't have to do it anymore. You know, I know you understand where I'm coming from.

> MS - Right, and seeing as you've raised this issue, I mean, can you maybe explain how these missions have taken a toll on your health?

Figure 11. Senator Frank Church holds up the "heart attack gun."

JP - Yeah, I'm forgetting a lot, Doc. I don't know if they're using a type of energy, type of technology, or vibration to my brain or to my body. My vocabulary is even more poor than when I started in the military.... I know people have noticed and all that. I do know three languages and all that, but my vocabulary is really bad. Sometimes, I forget what to do and how to say it, and at times, I forget a lot. So that's another thing that's bothering me. Bone shrinkage, there's like, if my bone is aging more than regular people, my eyes, they're getting worse, I'm getting more blind. It takes a toll, it takes a toll on your body, physically and mentally.

MS - And I guess people will ask, well, why don't they allow you to use some of the regeneration technologies there, given the toll it's taken on your body, as opposed to just letting you use conventional medicine?

JP - It's all a system, how they do things. Yeah, I asked myself the same question. Why can't I get something that can make me feel better? But I felt a healing sensation when Jafis gave me a drink the other day. I think that was like three, three weeks ago, two weeks ago. My body felt better than usual. I believe it was a type of medication they were giving me…. I felt peace, you know, drinking it and all that.

The blue drink JP received from Jafis was described in Chapter 3 and suggests that his health is being closely monitored by Nordics, who first made open contact with him in 2008. The Nordics were likely involved in earlier contact incidents dating back to when JP was only seven years old and living in upstate New York. In *US Army Insider Missions 1*, I explained that the Nordics and the USAF reached technology exchange agreements and that JP would be protected while being given access to advanced projects and allowed to disclose some of what he was witnessing to the world.[31]

MS - So it sounds like the Nordics do help with giving you some kind of advanced medical assistance, but that the military authorities that get you to go on these missions, they just send you off. It has these impacts on your body as well as the kind of mind control or the memory wipes and whatever things that they're exposing you to, take a toll on your physical health, and they don't really help you in any way other than just allow you to just go through the conventional medical facilities when you get back to your base.

JP - Yeah. I did feel like a growth after drinking this blue substance of type; it felt like juice, but it was more like a plant-based type of taste. I did feel muscle growth three days after drinking it. So, when I started working out again, I felt stronger. So yeah, I feel good now, but it takes a toll on people who do these missions, and it's rough. It's rough on all of us, you know? Then, just keeping this information in your head it messes you up, you know, about how society is right now. Knowing that we have the technology to heal everything, and having the [hidden advanced] energy that we have, and this technology not coming out, you know, it really puts a toll on you, a stressful toll ...

MS - It sounds like this is also a control mechanism as well. I think withholding these advanced medical technologies from people in the programs that it's a way of controlling them as well.

JP - Yeah, it could be. It's something that when you think about it, Doc, and you're doing these missions, you're asking yourself, "Hey, you know, why am I here? Why do they want this information? How am I going to benefit from this mission or this information I'm getting by going to the ark and bringing it back to other people who are higher than me?" You know, but what I'm benefiting from is now telling you [and] the public about this.

I think this is the reason I'm still doing this is because they're letting me give this information to you and keep giving me the green light. So that's the reason, but I don't really want to go through this. I don't really want to do these missions. You know, it's something that I think the public should know. And I think soon the public will know about not just what is a UFO or UAP and where they come from, but also the types of Inner Earth people that live on Earth or also the arks. All this information is going to come out little by little, drip-dropping it, you know.

Now they're saying that there's a Nazi base in the Bermuda Triangle that the United States was keeping a secret. I don't know if you heard about that, but they're going to try to tag that along with the arks. And I think people have to be careful what they connect. Well, I'm sure the Nazis were probably trying to investigate the arks because they kind of knew where the arks were with [the] maps that they had. Yeah, that's another crazy thing that's happening that a lot of people are talking about.

MS - Well, I know that most of the listeners of this update will be very grateful for you continuing to do these missions and giving these public updates. I'm sure they're very grateful and appreciate what you're doing and the sacrifice you're going through, given the toll it's taken on your health. And I certainly feel sorry that you're having to go through this. But I understand that ...

JP - And, you know, they take everything away from you.... I know it's sad, but they could take everything away. Also, you know, everything that you ever lived for, they could take everything away, you know, and leave you in the streets, you know. But, you know, it's for a good cause. I know a lot of people need to know the truth, and they need to know about all this that's happening.

MS - Yeah. Well, I'm grateful that you're still able to do these missions. And I'm grateful that there are people in the military who have the foresight to allow you to go public and give you the green light to go public. I mean, it would be great if they gave you the kind of advanced medical healing so that you're not suffering. But I guess we can't have everything, can we?

JP - No, we can't have everything. It's tough. I wish that. But there are other people involved. There are higher people [making the decisions], and we have to understand that. My father always told me that there's always a rooster stronger than another rooster.

MS - Yes, we do live in a hierarchical alpha-male society. So, JP, I didn't ask you at the beginning, but how many days ago did this mission happen?

JP - This happened Tuesday?

MS – Okay, Tuesday, so right now, it is Saturday, December 2nd. So, Tuesday that would have been November 28. Okay, so that's roughly the date you did this mission on. And you mentioned something about the Pacific Ark at some point. So, I mean, is there anything you wanna say about that now?

JP - I don't know if I can say anything about it now. I haven't gotten the green light, but I know the Chinese are involved with that ark. The Nordics took it over, and right now, they're trying to do the same thing that they're doing over here in the Atlantic. They are trying to activate it more than what it is. Yeah, I don't know if you can put that out. Yeah.

MS - Alright, well, I mean, that's enough. I mean, when you get the green light, you can tell us more about the Pacific Ark.

JP - Sure, it's similar to the ark over here in Atlantic. It looks the same. It's like cut and paste. You know, these arks are like cutting and pasting [as if] it was manufactured in the same place.

MS - Okay, so that Pacific Ark is somewhere in the Dragon's Triangle area there in the Sea of Japan between China, Taiwan and Japan, and it's a kind of similar ark. Okay, very good. Yeah. So, do you want to say anything to the Spanish and Portuguese listeners of the show? I think you've got a big following there, do you want to say something to them?

JP - I'm going to talk to the Spanish people first, then I'm going to speak in Portuguese. To all of you who are in Brazil. Thank you very much. I love you very much. I know you listen to Dr. Michael's program a lot, and it's very nice for all of you. The support you give us is very nice. It's very good. So, if you want more information about the missions that I have had, [visit] exopolitics.org. [There] you can see more missions that I did. A lot of love. Thank you. The day is approaching for disclosure. Yeah, basically, I told them if they need more information about the missions and all that about the stories to go to your Exopolitics.org. Yeah, they can see more information and maybe, in the future, translate some things for them. Yeah, that would be cool. That would be awesome.

MS - Okay, well, I want to thank you again, JP, for doing these missions and sharing them with us. We all appreciate what you're doing, and our prayers and love go to you and your family so that you are all safe and prosper in these challenging times.

JP - Yeah, I want to wish everybody a Happy New Year, Merry Christmas, and to think positively. Treat people beautifully. Treat people nicely. Man. Just think positively about every situation in real life, and you'll be blessed, all right! Yeah, I just wanna leave that with everybody. That's awesome. I love you guys.

MS - Thanks, JP.

JP - No problem, doc. Have a beautiful day. Bye-bye.

It's important to appreciate the significance of JP's statements that he did not wish to continue doing covert missions due to the health toll these were having on his body. Moving into and out of different dimensional frequencies, electromagnetic frequencies, gravitational zones, etc., were all exerting a big health toll on him during missions. Conventional medicine used in the Army was not helping him, and the assistance from the Nordics was not enough. The cumulative stress and toll placed on the body when performing his missions is discussed in greater detail in Chapters 11 and 12.

Chapter 5

Strange Medical Experiments & Atlantic Ark Update

Strange Medical Experiments

In a December 30, 2023 update, JP describes a bizarre series of medical experiments he underwent at a military hospital he was sent to along with five other soldiers. He describes the group of six soldiers being connected to unusual equipment and taking medications that led to very strange physical and temporal situations. He describes being hooked up to a brainwave scanning machine where his dreams would be monitored and how space-time was distorted in ways that tested one's sense of reality. JP says that the testing involved him and the other soldiers going to sleep and being repeatedly woken together at 2 am to test their reactions to unusual temporal distortions and situations. He says that a Nordic extraterrestrial was involved in running the experiments. JP was able to take a photo and video of the hospital ward where the experiments were conducted.

  JP also reveals more information about his prior encounter with Jafis, a Nordic ET, and his last Atlantic Space Ark mission with Nordics in response to concerns over whether he was being manipulated. JP describes what he felt about Jafis, who invited him to enter his spacecraft and the Nordics aboard the Atlantic Space Ark that had taken over temporary control from the Earth Alliance. JP explains more about the nature of the cooperative relationship between the Nordics he has encountered with the Earth Alliance. The transcript of the interview follows, with grammatical corrections and my commentary.[32]

Interview Transcript with Commentary

Key: MS – Michael Salla; JP – Pseudonym for US Army Soldier

MS - It is December 30 [2023], and I am with JP. He has a short update to share with us. So welcome JP to Exopolitics Today.

JP - Hey, Dr. Salla, I'm proud to be here to bring you this information that I got the green light on. And it's going to be interesting. It's really interesting. I'm getting flashbacks of what happened, and I could start telling you the experience. It was a weird experience. And I haven't had an experience like this in a while. I had an appointment, right, to go visit the hospital of this particular military base.

So, I was driving over there, and I saw that I was being followed, right? I'm like, oh man, okay. So, I'm going to go into the base. They're not going to go into the base because you show your ID to go into the base, and you know, whoever doesn't have an ID can't go behind you. I passed by the gate. I showed my ID, and then the car was behind me. They followed me for about 20 minutes. And I saw that they entered also and kept following me. I was like, oh shit. All right. They're following me. Okay, cool. So, I went to the hospital. I parked. I got my book bag and all that, but I felt I felt a sensation of doing things without being asked to do [them], like not mind control, but I had the sensation of time passing a long time and me staying in the same position for a long time. That was quite weird when I was entering the hospital. When I was going up the stairs, I felt like I was going up the stairs for days. I'm like, man, when are these stairs going to stop? Jeez. You know, like I kept going up. That was a weird sensation when you felt that time was going slowly. So, I got to the place, and then there were six other soldiers, right? And there was another soldier, his name was Dan. I remember his name was Dan, and he's like, "Hey, do you know why we're here?" I'm like, "I kind of know why we're here, but I don't know if I can tell you." He's like, "Well, I'm here because they're going to do a study on us." I'm like, "Yeah, something like that. I read about something like that. They're going to do a study on us." And then the other guy that was sitting in the corner, he says, "Man, this is about those crazy missions." I'm like, "What are you talking about?" And he's like, "Don't worry about it, man. Don't worry about it."

Everybody's hesitant to talk about their missions because I know there are other soldiers having different kinds of missions that are not connected to the missions that I do, and other soldiers do. So, we were waiting, and then the nurse was like, "Okay, everybody come to the back." So, we got in line, and we started filling out this paperwork. Basically, the paperwork was like a psychological evaluation. "Oh, do you feel good today? And how far was your house? Do you feel safe in your house?" You know, like different types of weird questions. Then we went into this hallway and we walked by these rooms, which said, Room 1, Room 2, Room 3, Room 4, Room 5, and Room 6. I was in the 6th room.

So, we all sit down and ... each person had a nurse, and we all start getting hooked up with these different wires all around our body. And I asked the nurse, oh, like, "What type of test is this, and how's that [done]?" [She replied] "Oh, it's a type of sleep study test dream to see where your mind is at, to see what's going on." So, they portrayed it like a sleep test, right? And they have an intercom, right? They talked to all the soldiers, "When you guys are ready, let us know if you need anything."

So, everybody was getting hooked up and all that. And then at two o'clock in the morning, we heard a big sound like a big horn. Like the horn, it was like, BAAAHHH! But it was loud, so I woke up scared as shit. I'm like, "What's happening right now!" We were all sleeping and we woke up, and we all started lining up. I felt like I was in basic training again. And some doctor, he was really rough. He's like, "Everybody line up right now, line up!" It felt like he was more like a marine [instructor] than an army instructor. And we just clicked. We were like, "Yes, sir!"

Like we all clicked, all these six soldiers that were there. We listened to everything that he was telling us to do. He's like, "Line up, we're heading down." So, we grabbed our stuff at two o'clock, and they took us to a room where there were a lot of seats, like a waiting room, and we sat down. There was another dude there at the front desk. He had dreads, and he was like, "Okay, you sit in the corner. You failed." He started saying that to each one of us. And then he even told me that I failed. [He said], "Okay, we're going to do it again. All of you failed." And everybody was like, "weird". Like, "What the heck is going on?"

So, we all ended up back in the room, and then it was two o'clock in the morning. Again, we woke up, heard the noise, and went back to the room. It was the same scenario. The doctor came in, "Everybody get in line." Boom. We got in line again. And then we started lining up. Then the doctor said again, "Follow me." So we follow him again. And then the guy tapped at me. Dan, he was like, "Hey, we, we just did this shit. What the hell is going on?" I'm like, "I don't know. Just keep doing it. Just, just keep following me." So, we recognized that we did it before. It was like a deja vu.

So, I was like, "Oh man. Okay. We're going to go to the waiting room. Okay. We can go to sit down, and we're going to see the same guy with the dreads. And he's going to ask us." The second time around, one person passed. The guy behind the counter with the dreads. He was like, "No, no, no. Everybody needs to pass. This is bullshit. All right. Let's do it again." Boom! So we go like robots, not robots, but we are not controlled, but we know what to do. We go back, and you can still see the nurses and other people walking around. Then I noticed in the corner of my eye Nordic-looking beings working outside. I looked at them, and they looked at me back, and we made deep eye contact. And I looked at him, and I remember his face. So, two o 'clock again, it happened the third time.

MS - JP, when you say two o'clock again, are we talking about consecutive nights, like this is the next night?

JP - Well, for me, it felt like the same day, like everything has happened back-to-back. So, we get on the bed, we close our eyes, and [then] open our eyes. There is the same thing. I think it could be multiple days, but I think they were putting the time back. And because we're in the hospital, we didn't know the day or nighttime. Everything was happening. So, it was weird. The third time it happened again, two o'clock in the morning, boom, the doctor shows up, and everybody gets in line. And I noticed when I was walking down, Dan, he tapped me. Hey, he said the same thing. I'm like, "Danny, we're, you know, we're doing the same shit again." He's like, "Yeah, man, it's like the fourth time." But I only remember the third time I'm like, "Oh my God, what's happening?" And then we all pass. We sit down, and then we all pass whatever will happen. And then they took us to another room. They connected multiple stuff in our hands, in our legs, and arms ...

At this stage, it's worth speculating about the purpose of the experiments JP and the other soldiers were being subjected to that involved medical monitoring and bizarre nighttime events. The experiments were designed to induce confusion and disorientation among the participants and see how each soldier would respond. All the soldiers were previously involved in highly classified missions, similar to what JP has been revealing during the four years of his Army service to date. Therefore, it appears that the experiments were designed to fulfill several purposes.

First, to test and evaluate the soldiers when put in confusing and disorienting situations often associated with the "high strangeness" of the UFO phenomenon. This would be important when they were sent on classified missions to off-planet or Inner Earth locations, where there were significant time dilation effects, as JP previously experienced in the space arks. Second, it creates a veneer of confusion and disorientation when it comes to soldiers discussing highly classified missions between themselves. Such a purpose would help maintain the operational security of classified programs and ensure they remain compartmented without significant cross-talk among participants. Finally, get the soldiers to question their own sanity in "high strangeness" situations. Again, this would suggest the experiments were designed to prepare for unusual classified missions with significant temporal effects and as an operational security measure for soldiers who might otherwise talk about their experiences to others outside of their military chain of command.

MS - You sent me a photo of all these electrodes connected to your head. Was that part of what happened?

JP - That was part, yeah, that was during the evaluation, whatever was happening. So, I had my phone with me. I wasn't supposed to. I think I wasn't supposed to have my phone. Cause you have to put [it away] like, it's like basic training, you have to put your phones away. But yeah, I said, "[how] do you prove that I was all connected?" You know. So I was like, "Oh man, okay." And I felt super tired. And then we all woke up, and it was three-ish in the morning. We all lined up. And then we went down this big hallway of the hospital. And then we went down the ramp.

It was like a maze. I tried to remember every part of it. And then we went through stairs that went probably five stories down. And we lined up, and I was like, "Hey!" I was only close with Dan. He was the only one who was talking to me. I said, "Dan, what do you think is going on now?" He's like, "I don't know, but I need my fricking shoes on." We were without shoes. We didn't have them. We were lined up, and there was a podium. And then we sat down in this big auditorium-looking place. There was starting [an event with] other soldiers [from] different branches coming in, but they were dressed. There were others with PJs on and others with regular shorts like us, some with shoes, some without shoes.

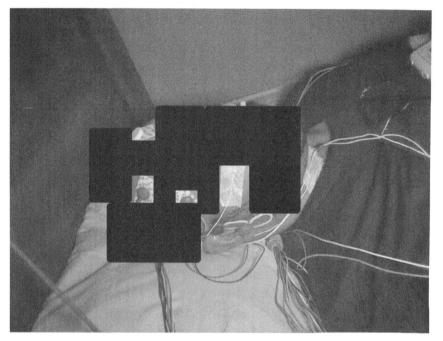
Figure 12 JP took a selfie while hooked up to electrodes during the medical experiment

JP - There was a holographic looking person talking on the podium. You could see that he was holographic because you could see through him. And then he was saying, "Okay, there's going to be a great mission coming up in 2024. We can't tell you the particular date, but this mass mission is going to be involving different ET races." He was well-dressed, you know. So, I was thinking, "Man, can this be like a video, or is this a real person projecting from another place." So, he kept talking about a mission that was going to go up. And I was just dozing off. I was trying to be polite to stay awake, and I was just dozing off. I dozed off to sleep, and I woke up and ended up back in the room.

It's likely that the upcoming mission in 2024 that JP was to be sent on was to Saturn, which I discuss in Chapter 11. It had a number of highly unusual aspects that might explain the bizarre medical experiments JP was involved in during this current mission. It's possible that the officer conducting the briefing was doing so from an off-world location close to where the upcoming 2024 mission would be conducted. When it comes to the use of holographic projections for conducting events, it has grown in popularity since it was first demonstrated in 2017 by Professor Stephen Hawking, who delivered a presentation in Hong Kong[33]. In the following photo from the event, one can observe the see-through effect that JP described happening at the medical auditorium. It's certain that classified versions of holographic technologies are far more advanced than what is available in the public arena. It's, therefore very possible that JP experienced a military officer at an off-world location using holographic technologies to give a briefing to soldiers assembled at the medical auditorium about an upcoming mission, possibly to Saturn.

JP picked up what happened after he fell asleep during the briefing and woke up again in Room 6.

> JP – [Woke up the] same time, 1:58, before two o'clock and I was waiting.
>
> MS - Just to be clear here, you're out there listening to this kind of holographic presentation by someone, maybe a computer-generated image, talking about upcoming missions involving different ET groups. You doze off listening to this presentation, and then you wake up and you're in the bed in the hospital room again.

Figure 13. Stephen Hawking speaking via Hologram in Hong Kong

JP - Yes, we were in a hospital bedroom, but it was like 1:58 [am]. So, I knew that the 2 o'clock [wake-up call] was going to happen. So 2 o'clock happened, no noise. Number five was there, and the rooms were connected by bathrooms, so I knocked. There's a room, bathroom, room, bathroom. However, each bathroom has two doors that each room can share. So everybody's connected. So I opened the bathroom. I knock on Dan's room. Dan opened the room. He looks at me with his eyes open. He's like, "What the hell, man? Did you experience what happened to me, too?" I'm like, "Yeah, bro. like, did you doze off, too?" He's like, "Yeah, I just felt freaking tired. I couldn't hear no more." I'm like, "Do the other guys know what's happening?" He's like, "No, I tried to knock on their door, but they're knocked out. So it's only you and me." I'm like, Dan, he's like a really big dude. He's like really strong. So he's like, "You know what, I'm going to start asking questions here." Like, "Bro, calm down. You went through this before." Apparently, he and I went through this before." [I say] "So we got to just try to remember what we heard and how we came back here."

He's like, man, "I don't know what's going on, but this is bullshit." I'm like, "I agree. This is bull crap. What's happening to us?" So it was like 2:15-ish. The nurse opened Dan's door, and she saw him talking to me. She's like, "Aren't you supposed to be sleeping?" And Dan's like, "Yeah, I was just asking questions because we're all connected." We're like, "We're all connected." And Dan was like, "We'll talk tomorrow, bro. We'll talk tomorrow." And then we closed the door. I slept, and we woke up the next day. I knocked on Dan's room, and the room was clean. No, Dan. And everything is nice and neat as if nobody was there. And I asked the nurse, "Wasn't there like five other guys there with me?" She's like, "No, it was only you. You're the only one that was doing the stuff."... I look [up] at myself, and then I go to the bathroom to brush my teeth. Then I decided to look in the garbage can, and I saw an extra razor blade there and a couple more papers that I knew I had not put in.

I called the nurse ... I asked the nurse, "If nobody was here, why is this in the garbage can?" She comes really fast and grabs the paper and grabs the garbage. So I know that was like from the other night.... So, I go back to the counter, and then I get my stuff. I sign some paper, and they're like, "Oh, how was the study?" So they actually [require us] to sign these papers and to see how the studies were, and I wrote it down in the comments and said, "It was weird. It was strange." And the nurse from behind the counter started laughing. And I noticed she had two patches in the back and had a lot of hair, and it sucked trying to get out.

I started heading back to the car, and I was driving away from the hospital. And then I remember that I forgot my dog tags. Like, I mean, I can't be without my dog tags, and I had a gold chain. So, I turn around, I freaking go back. I say, "Hey, I left my dog tags." I go back to the room, close my eyes, and swear to you, it was the last time I was all connected again. That's when I woke up, and I sent you the picture. I was confused, so I wanted to register this. Then I noticed it was 12 something or 11 something pm. I go to the next door, I knock on the door, and Dan is there. And he comes out with his eyes open. He's like, "This is bullshit. This is crazy. I'm getting the 'f' out of here." I say, "Okay, let's go. We'll go together." So, we go out together and the nurse was like, "No, you guys need to go back." The test is not done yet. I said, "No, we're going out." So, I go out with Dan. He's like, "Man, I knocked on the door, and you were not there. I was by myself." I was like, "shit, that happened to me too." He's like, "Man, this is freaking insane. I'm not coming here anymore. What the hell is going on?" I'm like, Dan, "What are you?" He's like, "I'm from the Air Force." I'm like, "Oh, okay. I'm from the Army." He was like, "Oh, shit, oh, man." So, he was pissed off. He's a big dude. Like, nobody can stop Dan. He's really built and all that, like twice bigger than me.

And I'm like, "Dan, calm down. I know you're pissed off." And then he comes ... and pulls the drip from his mouth. He's like, "Eff this. I'm getting the Eff out of here." I said, "Okay, I'm going to follow you." But then I followed him back, and he left in his car. He had a red Toyota. I remember that. He left, and I never got his number. I'm like, "shit, man." Dan. I remember his name, Dan, and he left. He was just pissed off. And I left too. So that experience was really, really weird. They called a couple of times, asking me to come back because I had left my dog tags. So, I'm like, "I'm going to leave dog tags over there. I'll pick it up on another day." So, not too long [later], I just picked them up, and it was normal. They had it in the bag, and I picked up the dog tags. It happened a total of five times. The same scenario, like if they were probably trying to make us insane, I don't know. It was weird. It was crazy.

MS - And you said that there was a Nordic being involved in the experiments that was observing you guys?

JP - There was a Nordic. I remember making eye contact with this Nordic being three times as I'm going through the same situation.

JP's response suggests that the Nordic was monitoring JP and the other soldier's reactions to the strange medical experiment being repeated on at least three occasions. This indicates that the experiment was being done to test for situations that might be encountered in an upcoming off-world mission involving the Nordics.

MS - You never got any clarity or any information about what the tests were for.

JP - I remember going into that auditorium and hearing about the next mission that we're going to have in 2024, but I saw a lot of different people there, a mixture of civilians, a mixture of military people sitting down, hearing this holographic person talking about these missions. He was well-dressed. The hologram looked so real, the holographic image. He was behind the podium. So, I don't know if the lights were shining from the sides. There were two lights shining from the sides, and you could see a light shining from the top. It was making a triangular shape, but in the middle, you could see the hologram-looking person. I remember that. I remember that happened once. But yeah, I remember this guy, Dan, being really pissed off and me going away with him to the parking lot to leave the hospital because it was weird. Only he and I remembered the whole situation. The other four soldiers, I never really talked to them or none of that, because I was [Room] 6 and Dan was 5. Yeah, it was a weird experience. I haven't experienced that since.

MS - You had no senior officer or anyone saying anything to you about these tests? I mean, how did you end up going to that hospital? Why did you go to that hospital?

JP - Because I was scheduled to go there, and you have to go to these appointments, you know. You can't just dismiss these appointments. You can't.

MS - It was scheduled. I mean, are these ... written orders you receive?

JP — No, they're not written orders. They're like appointments for something else that's happening with me, which I think you know.

JP has confided in me some of the medical issues he has been having related to his missions that have caused him to be hospitalized on several occasions. So, he is referring to him being notified in the same way as regular medical appointments.

MS - Okay, I see.

JP - So you have to go through these appointments for them to evaluate you.

MS - Okay, so some kind of psychological evaluation.

JP - So I got in the car heading back to where I live, and that same vehicle unparked and went behind me and followed me for 20 minutes. And I looked at it, and every time I slowed down, it slowed down. So, just making sure where I was going, making sure I was going back home, I guess, and making sure I was going to the appointments. So, I was kind of confused if it was the white hats or the black hats [following me]. So, this felt more like a negative experience than a positive experience for me because of all the confusion and all the weird things that were happening without you knowing. So when you don't know what's happening around you, you feel uneasy, and you feel being used, you know?

So yeah, you should have seen the eyes on this big dude when he was pissed off. Dan, he was pissed off. And I think he has been through a lot of shit too. I think a lot of people are soon going to be coming out, especially in 2024, about their experiences in the realm that I'm in, but there are good experiences, Doc, and there are bad experiences.

It's just the people that are, I guess, in charge of these experiences [that decide].

MS - Well, we know from the past that there's been a negative faction that doesn't like what you're doing, and you felt that this was a negative experience. So that raises the question, is this kind of like a setup to try and have you leave the military or being forced to leave the military?

JP – Probably, I think if a regular person goes through this, they will become insane. They will begin questioning. They will question their life, you know, like, "What's your purpose, what's happening," this and that. I remember the Nordic looking straight at me with his eyes, and we locked, but I did not hear any language from him.

MS - Did you feel he was friendly? Was he neutral?

JP - It was neutral. I saw him like working there, you know, that he had a different uniform, a uniform that is familiar to what I saw on the donut-shape ship in the Atlantic. So the doctor, he was an officer who was telling us to line up. The nurse had a weird, different uniform. It looked similar to a uniform from Star Trek. But it was a different type of uniform. But the last time we saw the nurse, she had the regular nurse's uniform. So, it was a lot of weird things that happened that was kind of confusing to me.

MS - Okay, and as you said, that was probably intentional, that they were trying to do something to stress you guys, to put you under some kind of mental pressure. It was part of a test, and I know you said to make you insane, that others would feel insane. Yeah, I think if other people went through this, and I'm sure, oh my God, I'm sure there's a lot of people that go through this type of situation, but I don't really think it's connected as much as to the military. I think it's connected to other people.

JP appears to be referring to a negative faction he has encountered many times during his covert missions, who are outside of the military chain of command. This negative faction includes individuals working for secret societies, such as the Freemasons, who operate independently of the wishes of the military. For example, in an earlier mission to an underground Ant Civilization in Florida in September 2022, gifts of different seeds associated with a Tree of Life and a sleeping giant were taken out of the US to England and given to the newly installed British monarch.[34] JP believes this negative faction was behind what he experienced in the bizarre medical experiments, as the following exchange clarifies.

> MS - I see. So you've done a lot of missions, we've reported on those missions. Typically, these are missions that have a purpose, and we've been able to talk about it, and you feel that you're working with white hats. But in this incident, you feel that this was something arranged by a negative faction, black hats, to put you under some kind of mental psychological strain.
>
> JP - Probably to grab information about certain missions because even though I go through these missions, there is a mixture of white hats and black hats involved. You know. But it's like politics, right? Democrats and Republicans. It's insane. And then there's the push of higher rank and lower rank and lower enlisted and higher rank. And you have to

comply. You know, if you don't comply with these higher rank people, you're screwed. So, you just have to mechanically do whatever they tell you to do, not unless one or two starts questioning like how Dan was doing. It was funny the way he was acting and quite surprising how he just got up and left. And I just followed him, and I said, "This is bullshit, you know, let's go!" Because we felt the same thing, you know, it's not like it was different. We knew that something fishy was going on....

Then we drove off. I actually drove to another base because, from there, we had to keep working. So, I didn't go home. I kept going to another base for me to keep working. I was sleepy, tired, and all that. They noticed that. So, because of New Year's and the holidays, it's like a half day, you know, and three-quarters of the day we're working and trying to get stuff done. So, one of the persons in charge saw that I was sleeping, and they … knew where I was. So, they sent me home saying, "Oh yeah, you can go. We got you covered." So, I left; I went home after I went to the base. It was quite an experience, you know. It wasn't a mission and all that, but it was an experience that I think other people go through, and it's rough.

More on Nordics Taking Over the Space Arks

MS - Okay, well, I want to ask you some questions about two of the previous missions because there's been some controversy over whether or not the Nordics involved in those missions are positive or not. So I just wanted to go back to the mission or the experience you had with, I think it was Jafis where you got out of a car and you went into the bush and

Jafis gave you a blue liquid, and I just wanted you to clarify whether you feel overall that was a positive experience, whether Jafis was in some way compelling you mentally. Yeah, both your general impressions of what you experienced with Jafis in that earlier update you gave me (see Chapter 3).

JP - I felt safe. I felt that he was trying to help me in a way. Yeah. I don't think I felt that negative vibe coming from Jafis. The ship looked familiar to me. You know, it looked like that ship. I took a picture of the ship [in Orlando in 2018].[35] Well, I felt peace. I was a little bit hesitant about the liquid, a little bit, but he knew I was hesitant. He said, "No, it's safe to drink. This is to help you." So, I drank the liquid. Yeah, I think I felt more peace than a negative feeling with Jafis.

MS - Did you feel any kind of compulsion during the whole experience because you left your car on the side of the road? When you parked it, were you under some kind of mental compulsion to stop your car, get out, and go to the craft, or did you just see the craft, stop the car, and park it?

JP - But it wasn't like a robotic feeling, but I knew I was being fished. When I say being fished, you know, when you put a worm on a hook and throw it into the water, you know, you're going to catch a fish. So when I see a ship landing right beside me, that's my worm. I'm going to get out of that car and I'm going to check it out. So I think I was fished, yeah, but not controlled.

MS - ... Okay, so it was free will all the way.

> JP - Yeah, it was free will, but yes, I was fished. I was lured to go to the ship to check it out because they knew I like to check things out. I guess that's what happened in Brazil. I was fished.

The critical difference here between the contact experience JP is describing and the abduction experiences reported by UFO researchers such as Dr. David Jacobs and Budd Hopkins is that the former experiences are voluntary while the latter are involuntary.[36] Contactees are invited into the craft, and it's a free will choice they make, which is respected. Nevertheless, as JP notes, it can be said that the contactees are being lured into the craft by the extraterrestrials stimulating a human's natural curiosity and are therefore being "fished," as JP puts it, but it is still a free will choice. Abductees, however, are given no such choice. They are typically taken while asleep, and therefore, no permission is asked, and their free will is violated. When awake, the abductees are routinely taken despite any opposition they have. In the rare cases where abductees successfully resist being taken, it's because they made a determined free-will choice or have invoked a celestial entity such as Jesus, or another extraterrestrial group intervenes to help them, as occurred in the case of Elena Danaan.[37] Overall, the alien abduction experience is not a free will choice, while extraterrestrial contact is the opposite. Unfortunately, the 'experiencer' term that is increasingly used by modern UFO researchers blurs this distinction.

> JP - I saw the light, and they kind of knew this guy was going to come out, and he was going to check us out, you know. I went and touched the ship, and the rest is history. I think they know who they can fish and what type of fish they're going to catch. They're really careful of what lure to use to bring you to the experience of Interacting with ETs... It was an interesting experience, but yeah, I did not feel any fear. I did not feel sad. I did not feel what I

felt in the hospital, you know, like that negative not remembering type of feeling that they're using me, you know. It was a more pleasant experience with Yasif.

MS - I see. With the [Atlantic] Space Ark, the sixth mission, where you went down there and there were a bunch of Nordics that had taken over control of the Space Ark. Again, there's some concern over whether these Nordics are positive or not. So, what was your feeling interacting with those Nordics down there?

JP - Well, when I was in the ship, there were more things activated there.... I had a beautiful feeling that everything was working out. And you look at their faces, and they're smiling. All the ones are really serious because they're working. But yeah, I did not feel anything negative. But looking at the other people who went with me, they were kind of scared and surprised at how fast they [the Nordics] activated all the things on the ship. Like how do they know this? Wow. Yeah, they are hundreds of thousands of years more advanced than us. So, I suppose they know a lot of things ... they should know a lot more things than us.

MS - Well, there's been some concern expressed that the Nordics needed us to gain access to the Arcs, but once they gained access, they're now taking over control, and they're doing this. So did they need access from humans, from the normal surface humanity, or were they ...

JP - They still do need access because of us. We're involved. We're part of what they're working on and what we're working on. It's an asset. We found it [the ark]. These arks are really important to humanity. So we still have control. It's not like we stopped having control in a way, but the Nordics, they have control of the ship itself. Like, man, how can I [correctly] say this in English? Okay, ... they have control over the situation of how the ark is going to be turned on because they're like scared of how we're going to do it and who's going to be in charge of it. So having the Nordics in charge is going to be neutral for everybody involved, for all the other nations, you know. I remember they moved the ark more into the Atlantic so all the nations could have access to these arks, not just the United States, or not just China, or not just Russia, or not just Brazil. Everybody is going to be involved. It's the whole humanity that is going to be involved with these arks. So when they saw that one was really in charge of everything, they were like, okay, no, this is for everybody. It's not just for one particular people [nation], you know. So, I believe personally that the Nordics are just scared of how things are going to be, just with us in charge of the ark.

JP is here raising a critical point with the discovery of ancient technologies, such as space arks. To whom do they belong, and who controls access? International law tells us that such ancient discoveries belong to the country in whose sovereign territory they were first discovered. However, in the case of the Atlantic Space Ark, it was found somewhere in the Bermuda Triangle, presumably outside the 12 nautical miles that make up the territorial sea of any major nation such as the US, but perhaps close to the territorial sea of a small island nation. According to the 1982 United Nations Convention on the Law of the Sea (UNCLOS):

> *Article 2 – Legal status of the territorial sea, of the air space over the territorial sea and of its bed and subsoil.*
> 1. The sovereignty of a coastal State extends, beyond its land territory and internal waters and, in the case of an archipelagic State, its archipelagic waters, to an adjacent belt of sea, described as the territorial sea.
> 2. This sovereignty extends to the air space over the territorial sea as well as to its bed and subsoil...
> *Article 3 – Breadth of the territorial sea.*
> Every State has the right to establish the breadth of its territorial sea up to a limit not exceeding 12 nautical miles, measured from baselines determined in accordance with this Convention...
> *Article 6 – Reefs*
> In the case of islands situated on atolls or of islands having fringing reefs, the baseline for measuring the breadth of the territorial sea is the seaward low-water line of the reef, as shown by the appropriate symbol on charts officially recognized by the coastal State.[38]

According to UNCLOS, if the Atlantic Space Ark was found within 12 nautical miles of an island nation, then that nation would have sovereignty over the ark.

A more likely possibility is that the Atlantic Space Ark was found within the 200 nautical miles (370 km) that make up the Exclusive Economic Zone of any country adjoining the Bermuda Triangle. According to the 1982 United Nations Convention on the Law of the Sea (UNCLOS), an Exclusive Economic Zone (EEZ) is:

> [A]n area of the ocean extending up to 200 nautical miles (370 km) immediately offshore from a country's land coast in which that country retains exclusive rights to the exploration and exploitation of natural resources.[39]

However, it's doubtful that an ancient space ark would qualify as a natural resource that can be explored or exploited even if found within the EEZ of any country or island.

The most likely scenario is that the Space Ark was found outside of the territory of the sea and the EEZ of any country, in what's regarded as "the territory of the high seas," which UNCLOS declares:

> *Article 87 – Freedom of the High Seas*
> The high seas are open to all States, whether coastal or land-locked. Freedom of the high seas ... comprises, inter alia, both for coastal and land-locked States:
> (a) freedom of navigation;
> (b) freedom of overflight ...
> (d) freedom to construct artificial islands and other installations permitted under international law, subject to Part VI ...
> (f) freedom of scientific research, subject to Parts VI and XIII.
> 2. These freedoms shall be exercised by all States with due regard for the interests of other States in their exercise of the freedom of the high seas, and also with due regard for the rights under this Convention with respect to activities in the Area.[40]

According to UNCLOS, if the Atlantic (and Pacific) Space Ark is located in the High Sea, then it is considered accessible to all countries under UNCLOS, which has been ratified by 169 states, including major sea-faring nations such as Britain, China, and Russia, but critically, not the US.

Figure 14. List of Parties to United Nations Convention on the Law of the Sea. Source: Wikimedia Commons

What we do know with some certainty is that the US Navy has been exploring or exploiting the Atlantic Space Ark since at least May 2016, when it placed a donut-shaped ocean platform above it and granted access to different international and extraterrestrial visitors, which is what JP reported seeing when taken there on his first visit.[41] When JP was taken five years later to the same ocean platform in January 2022—after he had joined the US Army—JP was granted access to the Atlantic Space Ark in a joint US and China multinational mission. Did the US Navy, having effective control over the Atlantic Space Ark, give the US the right to make decisions over who to grant access to it or assert sovereignty over it?

Under UNCLOS, the Atlantic Space Ark—the same principle applies to the Pacific Space Ark—would be widely considered part of the "common heritage of mankind." UNCLOS asserts:

Article 136 - Common heritage of mankind.

The Area and its resources are the common heritage of mankind.

Article 137

Legal status of the Area and its resources

1. No State shall claim or exercise sovereignty or sovereign rights over any part of the Area or its resources, nor shall any State or natural or juridical person appropriate any part thereof. No such claim or exercise of sovereignty or sovereign rights nor such appropriation shall be recognized.

2. All rights in the resources of the Area are vested in mankind as a whole, on whose behalf the Authority shall act. These resources are not subject to alienation.[42]

UNCLOS does not specifically refer to ancient, submerged objects in its discussion of the common heritage of mankind, but instead refers vaguely to "The Area" in which such resources are discovered. There is another international treaty that is more specific when it comes to international law governing control, access, and ownership of the Atlantic and Pacific Space Arks. This treaty is the Convention on the Protection of the Underwater Cultural Heritage, which entered into force on January 2, 2009, with 77 countries being parties to it. The Convention begins with definitions and states:

For the purposes of this Convention:

1. (a) "Underwater cultural heritage" means all traces of human existence having a cultural, historical, or archaeological character which have been partially or totally under water, periodically or continuously, for at least 100 years such as:

(i) sites, structures, buildings, artifacts and human remains, together with their archaeological and

natural context;
(ii) vessels, aircraft, other vehicles or any part thereof, their cargo or other contents, together with their archaeological and natural context; and
(iii) objects of prehistoric character.[43]

A submerged space ark lying at the bottom of the Atlantic or Pacific oceans, whether it lies within the territorial sea or Exclusive Economic Zone of any country or in the high seas, qualifies as a vessel under clauses 1(a) ii & iii of the Convention on the Protection of the Underwater Cultural Heritage. The Convention further states:

> 3. States Parties shall preserve underwater cultural heritage for the benefit of humanity in conformity with the provisions of this Convention.
>
> 4. States Parties shall, individually or jointly as appropriate, take all appropriate measures in conformity with this Convention and with international law that are necessary to protect underwater cultural heritage, using for this purpose the best practicable means at their disposal and in accordance with their capabilities.[44]

Most importantly, in terms of the space Ark being found inside the territorial waters of a nation, the Convention states:

> Article 7 – Underwater cultural heritage in internal waters, archipelagic waters and territorial sea
>
> 1. States Parties, in the exercise of their sovereignty, have the exclusive right to regulate and authorize activities directed at underwater cultural heritage in their internal waters, archipelagic waters and territorial sea.[45]

As far as the space ark being found in the EEZ of any country, the Convention states:

> 2. A State Party in whose exclusive economic zone or on whose continental shelf underwater cultural heritage is located has the right to prohibit or authorize any activity directed at such heritage to prevent interference with its sovereign rights or jurisdiction as provided for by international law including the United Nations Convention on the Law of the Sea."

Clearly, this provision would empower the US to control access to the Atlantic Space Ark if found in its EEZ or the EEZ of an allied country.

Many major maritime countries—Australia, Canada, China, India, Russia, the UK, and the USA—are not parties to the Underwater Cultural Heritage Convention, meaning that they are not subject to any international enforcement mechanisms associated with the Convention, but private parties are nevertheless subject to its prescriptions in participating nations according to international law.

Given the international legal complexities arising from the Atlantic Space Ark being located in the territorial sea or EEZ of any country, the Nordics and/or the Ark's organic consciousness decided to move it deeper into the mid-Atlantic Ocean, meaning that the Ark would now be unambiguously in the high seas. This would enact provisions in both UNCLOS and the Underwater Cultural Heritage Convention in dealing with control, access, and sovereignty over the Atlantic (and Pacific) Space Ark. My discussion with JP resumes over international legal issues concerning who has control over and access to the space ark.

> MS - The United States, if it's in control of the ark, the Nordics perceive that there might be some

misuse or that could lead to tension with other nations that feel left out.

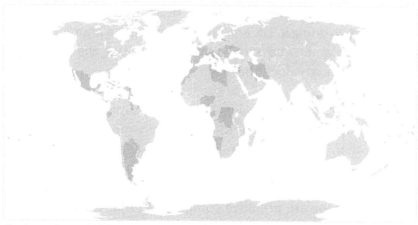

Parties to the UNESCO 2001 Convention on the Protection of the Underwater Cultural Heritage, as of February 2015

Figure 15 List of Parties to Underwater Cultural Heritage Convention. Source: Wikimedia Commons

JP - Not just Americans, because there are other people involved in investigating these arks. It's like a more world [involvement]. It's like certain religions [are involved]. You got the Inner Earth people also involved, and It's a present for us as much as it is for the Nordics, So the Nordics want to be involved as well. But because the Nordics are super more advanced than us, they have more access to the ... arks. So, they're as interested as we are interested, but they just have more technology to activate more stuff on the ship. It's like, okay ... you go to the Amazon, right, and you give a cell phone to an Indian guy, right, and he doesn't know what to do with it. But if you give it to somebody who is advanced, living in New York, and [they] come down to the Amazon, right, and they look at the cell phone, they know how to work with it. The [Amazonian] Indian guy is gonna look at that [New York] guy like, "Man, he's taking control of it, but he's showing me videos. He's showing me stuff with it. So cool. Let him just keep doing it. Let him just keep using the phone, and we can just be with him while he's using the phone and learn along the way."

That's a literal analogy, you know, similar to what's happening now. Nordics are really more advanced than us. And they're involved in a lot of different projects, not [just] this one, in our solar system and also in our oceans and also around Earth. We know that we are primitive compared to the Nordics. So, we're letting them [take control] ... And if the Nordics want to eliminate us in a certain way, they could. They have the technology [to eliminate us], but they don't. They're really not robotic, but they're really neat in how they do things. And they're really fair in how they do things. It's not like, you know, how can I put this? Do you understand what I'm trying to say using the cell phone analogy?

MS - I think the comparison is good that the Nordics have a superior understanding of the operation of the arks and by them taking control, but they're still working with the Earth Alliance, and they're showing the Earth Alliance scientists and military personnel how to better use the ark, but doing it in a way where it's not just one country in control, but it's more of an international effort.

JP - Yeah, and let me tell you something, a lot of people will feel uncomfortable ... because I bet you, like back to the analogy of the Indian Amazon, if they see that they [the Nordics] are not sharing information they're going to be a little bit scared [that] they [the Nordics] are holding back information that they think we're not ready for, you know. They're going to be quite scared, but the ark has so much knowledge of the Earth's history that it can affect religion. It can affect a lot of things on Earth. So, they're really hesitant to give out information on certain subjects that the ark carries, you know, certain historical memory that the ark carries that would change a lot on Earth, certain stuff, yeah.

MS - Overall then, you're not concerned by the Nordics taking over the arks, that you don't feel that this is something bad, that this is actually maybe a temporary measure to kind of...

JP - It's beneficial for everybody, for them [the Nordics], to have control now because they're unlocking certain stuff that we would have never unlocked. So, we're right now the Indians, just being with them and making sure that they share what they learned and making sure that they're transparent. That's what everybody's seeing with what they're learning about the ark. They [the Nordics] are being transparent, but they are not completely fully in charge. They still want us to be part of the story, you know. So yeah, that's what's happening.

MS - Okay, well, it's Saturday, December 30. It's one day before New Year's Eve. By the time this goes out, it'll be 2024. So, is there anything you want to say to people about 2024, what's coming, the space arks, or what do you think is going to happen?

JP - I believe 2024 is going to be a big, big deal with the UFO community and disclosure. A lot of people, a lot more people, are going to be coming out. A lot of whistleblowers are going to be coming out and talking about their experiences. There's going to be doctors, there's going to be pilots, and there's going to be scientists [coming out]. Even though that's happening now, you know, a lot of people are coming out now, but it's going to be [even] more people coming out, with more details, with more evidence. It's going to be insane... We do have to be careful because technology is evolving, and we have a lot of fake stuff going around ...

There's going to be a lot of people that are negative, and they want to stop all this info coming out, and they're trying their major best to have this information not to come out because we're not so-called "ready". But I believe that, with all this that happened in 2024, there are a lot of people who are more open and more accepting of everything that is going to come out in 2024. It's going to be exciting. I think 2024 is going to be exciting. These things that are happening with the arks. I think they're gonna try to block as much of it until these ships appear. Even though they're getting closer from Jupiter ... [from the] asteroid belt by Ceres. There's gonna be more information coming out on on the Moon, Phobos, that people are gonna be really excited about in 2024, the so-called monolith that is on Phobos.

The monolith on Phobos was first observed in a photo taken by NASA's Mars Reconnaissance Orbiter mission in 1998 and is very similar to another monolith-looking object photographed on Mars's surface.[46] What really brought attention to the mysterious monolith-looking object on Phobos was Apollo 11 astronaut Buzz Aldrin, who asserted:

> "We should visit the moons of Mars. There's a monolith there - a very unusual structure on this little potato-shaped object that goes around Mars once every seven hours. When people find out about that, they are going to say, 'Who put that there? Who put that there?'"[47]

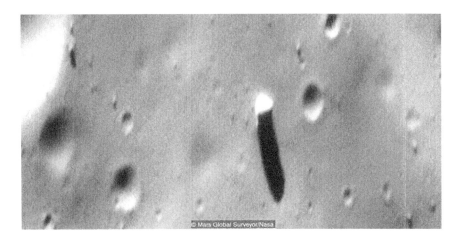

Figure 16. The Phobos monolith - Mars Global Surveyor/NASA

JP - There's gonna be a lot of information coming out in 2024 that is gonna shock a lot of people. More [information] about interdimensional travel, vibration, and frequency, and how we can use simple technology to levitate things. It's gonna be a game changer on the things that we're gonna find on the arks and in space, and different parts of our solar system. It's gonna be interesting.

MS - Sounds awesome. So any final words for our Spanish and Portuguese friends? ...

JP - And for everybody who is listening, I opened up an Instagram account. It is called JP.missions. 2024 is going to be amazing. There's going to be a lot of revelations. ... So it's going to be really exciting with everything that's happening. God bless everybody.

MS – Well, thank you, JP and I wish you and your family all the best for the New Year. I look forward to a big year of disclosures and, hopefully, more mission reports.

JP - Roger that Doc, I'm happy to be here.

Chapter 6

Nordic Assimilation Facility

On January 12, 2024, JP and three other soldiers were sent on a mission to meet with four Nordic extraterrestrials dressed in distinctive uniforms and with long hair. After their arrival, JP, the soldiers, and the Nordics traveled to a military facility where the four Nordics went inside. When they exited the facility, they had been physically transformed. Their hair was cut short; their clothes were now modern-looking jeans, shirts, and boots. Also, they now possessed official identity papers and passports. JP learned that his mission was to protect the Nordics from their arrival up to their departure when they were taken to an airport where they would travel to distant locations as normal humans and not stand out in any way.

JP's description of what can be described as a Nordic Extraterrestrial Assimilation program that is secretly being implemented by the military intelligence community is very similar to what contactees reported in the 1950s and 1960s, where they would help new Nordic ET arrivals assimilate into human society. This ad hoc assistance by the early contactees was now done in a far more comprehensive way due to secret agreements that had been reached between the Nordics, the US, and global leaders. The transcript of the interview follows, with grammatical corrections and my commentary.[48]

Interview Transcript with Commentary

Key: MS – Michael Salla; JP – Pseudonym for US Army Soldier

MS - We are here again with JP. It is Monday, January 15, and he has another update to share with us. Welcome, JP.

JP - How are you doing, Doc? It's a pleasure to be here and a pleasure to bring this information out. Got the green light; let's get on to this.

MS - For those who are still new to you, I just repeat again: I've known you since 2008, and you had contact experiences. You contacted me, and we continued to communicate. In 2017, you started sending me photographs of [antigravity] craft, triangle, and rectangle-shaped craft that you were seeing over or near MacDill Air Force Base in Tampa, Florida. In 2019, you joined the US Army, and you are doing missions with the secret space program or US Space Command, or whatever we want to call it while having regular duties within the Army. And you're being green-lighted. You get approval from your covert chain of command to talk about these missions, but your regular chain of command doesn't know anything about it, and they are not that happy. Is there anything you want to add to that for people who are new?

JP - Yeah, negative. You said it all.... It is what it is. When you're chosen, you're chosen, I guess. And I know a lot of people would agree with that. They're in a service, and they're chosen to do different things that overwhelm them. And it surprises them, and [they] never knew that It was in the realm of real life. And it shocks you when you go through these experiences, and they choose certain people, they fish certain people out. We go through these experiences, and, yeah, it's incredible the things that we experience. The [things] some people go through. Right.

MS - And we need to emphasize that you were chosen because the Nordic or these human-looking extraterrestrials that have Nordic features contacted you in 2008, and the US Air Force in particular and other agencies were tracking those contacts and tracking our communications, and they wanted you to join the military, which you did after many years of them encouraging you.

Regarding the technology to track extraterrestrial contacts, spacecraft emit large amounts of gamma wave electromagnetic energy when appearing in our atmosphere, and this allows them to be tracked anywhere in the world through satellite surveillance.[49] It's possible that this is how the intelligence community first learned of JP's contacts with Nordics, and it has been tracking him ever since.

MS - So, it's not just a matter of you now, all of a sudden, coming forward with all this information. There have been many years of preparation, of the Nordics working with you, you having contacts with them, you having contacts with the Air Force, and eventually, covert Air Force people convincing you to join the military so that you could be more tightly integrated into the world of covert operations because as a civilian, there were some roadblocks still.

JP - Oh yeah, there were roadblocks and all that, but they knew, I guess, that I know three languages. So, I guess I was a good candidate to bring out this type of information during the same time that everything else started going on with Space Force, with [David]

Grusch, with everything coming out at the same time your books were coming out. So, everything was synchronized in a way that was quite interesting. So everything is connecting with each other. So that's quite interesting. It's quite awesome what happened in Tampa Bay and Orlando, as well as this other base, and how everything is getting put together. And there are a lot of other people involved, but I'm kind of scared that sometimes misinformation is put out there from another part of ... how can I put this? Another part of the region. I just can't say certain words. So, I'm trying to be careful how I talk, Doc, how to bring out these certain words, because there are certain words I can't just say out of nowhere.

MS - Okay, well, let's get to this latest mission then. I mean, this is January 15 [2024]. When did you go on this mission that you want to give us an update about?

JP - The 12th, Friday.

MS - Okay, you want to walk us through what happened?

JP - So we went. They called us. We went to this particular base. We got into a van, and there were four guys. [Then] we were there, and I saw Dan again. Dan was there [and said]: "Ah, Is this the mission that they were talking about?" I'm like, "Bro, I don't know, but it's cool seeing you again." So he was there, and the two other guys had a type of mask on, so we didn't recognize them as much. But Dan did not have one. He had on just regular sunglasses and a cap ...

MS - Just to repeat, Dan was the soldier that was part of these medical experiments that you described in your last update [see previous chapter]. And he had the adjoining room to you, and you guys talked about what was going on with that whole, very bizarre medical experiment.

JP – Yeah. So, we were there, and we were talking about what happened in the hospital. He's like, "Yeah, that shit was weird." That was interesting. So, we were talking, and we went with this van to this field. And the field was an open field, and we heard in a distance a Black Hawk coming in, a UH-60. It was coming in, and it was hovering for a little bit in our location. I'm like, whoa, what the heck is going on? ... There was a helicopter hovering over the trees, getting ready to go to our location. But above the helicopter, there was a black type of ship, right? And it went over the helicopter, and. then it zipped up higher, close to the clouds, like, waiting, just parked there.

"Do you see that?" "Yeah, I freaking see that, that's interesting." So, we all started talking about it, and then the helicopter started coming down little by little and softly and softly. It was coming down, and it [the wind] was getting stronger. So, we lined up on the side, away from the helicopter and the wind. It was freaking cold, doc. It was amazingly cold where we're at. We were freezing. And I'm like, "Man, I need some gloves." So, Dan took some gloves from his book bag, and he's like, "Hey, here, you can use my gloves. But these are combat gloves. They're not warmer." When you wear combat gloves, your hands get even colder. I'm like, "Man, you don't have other ones?" He's like, "Bro, it's either this or nothing. I'm like, "All right." So I just put them on, and we're waiting here ...

We're dressed up ... in black military attire. The helicopter landed. And then, out of this Blackhawk helicopter, two pale-looking Nordic beings came out dressed in uniform. I'm like, "Man, they're coming out of a helicopter, Dan." And then Dan's like, "Hey, do you see this, bro?" I'm like, "Yeah, I see this." And then the guy, the other guy that was with us, one of the four guys, he was like, "Hey, tell them to get over here and get away from the helicopter."

So, we waved at the Nordic, pale people to come to us, and we were like, "Hey, over here, over here." They pointed at us, and then they started heading to us. They started not running but jogging to us. And their English was so perfect, like, "Hey, how are you guys doing? Oh, yeah, we just came from the freaking arks!" They started talking like us. Me and Dan looked at each other. We're confused. We corner [squint] our eyes and all that.

MS - Can you just clarify? When you say they start talking like us, you're saying they're talking with a New York accent or like an American ...An American accent from, I don't know, like Tennessee or, I don't know what accent it was, but it was a really American accent. And then me and Dan, we're just like, "What the heck?" This is freaking weird. But we looked up, and the helicopter was still landing. We couldn't see the pilot. It looked like this helicopter did not have a pilot because we tried to see in front who was flying the plane. Not the plane, the helicopter. And nobody was there. And I was like,

"Yeah, that was weird." The helicopter, it was like a drone type of helicopter, but it was a Black Hawk. So we were like, "Okay, that's new. That's nice.' So, the helicopter took off and flew out.

The two pale Nordics came into the van. They're like, "Wow, it's cold!" Put on the heater, and the heater just turned on in the van. The heater turned on. We kept the two Nordics inside the van, and then one of the Nordics said, "Oh, we're waiting for two more." The ship that was above the helicopter, the black-looking ship, and I sent you the pictures of these ships, right? I don't know if you got them.

But I did send you a picture of this ship. It looked like an oval rectangle ship, right? And it was really thin, and it started coming down slowly. We looked at it, and it was making a humming noise. The helicopter already left. It took off, and it left, and it made way for the ship to come down.

MS - JP, so as that ship was coming down, you saw it, and you took photos of it because that's what you sent me. You sent me four photos, and three of them show some kind of ship in it.

JP - In the first photo that I sent you, if you look at the top corner, you can see the ship. So, the photo is on the top, so you see something black in the corner of the picture. That's the ship itself, and then the other three is a ship leaving into the woods. But these pictures were not [taken] at that exact time. It was after we went back to the base. You can see it, like, coming in and then going out. So, this ship started landing, and it never touched the floor. It was making a humming noise, and we felt, like, a vibration all over our bodies.

Photos 2 & 3 in a sequence of 4 taken by JP on Jan 12.

Enlargement of saucer shaped craft on top right of each photo which 2 Nordics from the Moon

Figure 17. Photos of a craft similar to the one in which two Nordics from the Moon arrived

JP - And the Nordics inside the van, they were laughing at us because we were acting weird around the ship. And we've been around these things a lot, but every time you go through the experience, you're always overwhelmed. It's like winning the lottery. You're overwhelmed. Every time you see these beautiful, magnificent ships, you are overwhelmed. You are like, "Wow, this is beautiful. This is cool. This is awesome." So, you get so excited, but you can't just go crazy about it. You have to stay put and wait and do your part. So, the ship started coming down, and it hovered, like, 3 ft [0.9 meters] above the ground, right, and it was, like, the size of three or four minivans. Like, side to side, like that size and that height. And then the ship, it was, like 4 ft [1.2 meters above ground], and then it curved and touched the ground, and then on top, a little capsule opened up. It was a ship that I hadn't seen yet. Let's put it in perspective. Okay, the ship is 4 ft above ground, right? It turns at an angle, and part of the ship touches the floor. And then, on top of the ship, the side opens. Then, two Nordic beings dressed similarly to those that came out from the Black Hawk helicopter came out and started speaking a different language.

And it felt like it's an ancient Greek [dialect]. It sounded like Greek, but I could pick up a couple of words from Greek, but it was not Greek. I think it's an ancient type of language, but it's similar to the Greek dialect. So that was really interesting. I love to pick up on language. I love what music is because music is the oldest language we have on Earth, and I love languages, so I try to pick up on languages as much as I can. I'm trying to learn Russian, I'm trying to learn South Korean, I'm trying to learn Greek. … I'm trying to learn, like I said, Russian … So, I'm picking up on different languages all of a sudden, a lot of Portuguese …

I'm picking up different languages because the … same part of the brain you use for music, you also use for languages. So that's quite interesting. So, if you're really good at music, you're going to pick up on different languages … I started picking up their dialect, and I'm like, "Wow, cool. All right, awesome." So, they came, and then the Nordics inside the van said, "Oh, they're from the Moon." I'm like, "what"? And then me and Dan looked at [each other] and we opened our eyes [wide]. We're like, "Holy crap, they're from the Moon. What part of the Moon?" They did not say what part of the

Moon they were from and which moon they were talking about, so I'm thinking it's our Moon. And the Nordics, they never told us.

So, they came into the car, they started talking to the other Nordics that were there, and then the other Nordics started speaking to them in that language, and they were so happy to see each other. They haven't seen each other for a long time. Like if they're related or something like that ... or they're cousins, I don't know. But they seem so happy to see each other.

MS – ... Can you explain how the two Nordics that were already in the van, and the two that just came, that were from the Moon, how were they different? I mean, you describe them both as Nordics. What was the difference, apart from the language?

JP - The ship they came in and their uniform was slightly different, but they looked the same. They were pale. They looked the same as the Nordics, who came from the ark.

MS – Okay, and the uniforms were slightly different. I mean, are we talking about uniforms that are kind of similar to, say, Air Force uniforms?

JP - No, it's different. The Nordics have their own uniform. It's more tight. They have, like, their shoulder shirts are pointed out, and then they [come] in. Similar to the ones we see in Brazil. But it's smaller. It's smaller. The shoulder is not that big; it's smaller. But I don't know if it's due to their rank or something like that. The uniform is similar to ours as well.

And they're more tight on them. I think there is a protective part on the uniform that protects them from radiation in space. So, this uniform, when you look at It, and they turn it a little bit, side to side, it turns to different colors. Like, if it was a lizard underwater. When you put a lizard underwater, it grabs, like, a bubble, and when you turn it around, it has different colors of it. So, it's similar to that. I don't know if it does that because of our atmosphere, our oxygen level and atmosphere. But when you look at the uniform, and you turn around and look at it, it turns into a rainbow-ish, different color type of uniform. But the uniform itself is like a bluish, light blue color, and it's just built differently. The shoes are connected with the uniform. So, in order, I guess, to take the whole thing off, you have to take off your shoes too. It is what it is. There is a way of getting dressed. So it kind of looks cool and futuristic. And the Nordics, we were surprised by the way they were acting because they totally act like the ... Star Trek people, the Vulcans. They really act like that without any feelings and all that. So that's quite interesting.

MS - They behave like Vulcans?

JP - Yeah, they behave just like Vulcans. And I don't know how Star Trek got that so spot on, but the Nordics behave just like Vulcans. Yeah. The way they talk, the way they're so straight on, they're straightforward. But these guys that we picked up, they were more like us, and we did not understand that, so we took them. They came out from the ship, and they started walking to the van. And the ship just curved, backed up 4ft, the capsule closed up, and the ship, just in the blink of an eye, went straight up to the clouds. And you can see the indentation of the clouds when the ship flew up, and we were like, "holy crap, that was like more than 3000 mph." We started talking about the speed, and everybody started laughing about it. It's like, "holy shit, that was like really fast. Oh, that's what these pilots are talking about. Oh, yeah." Okay, so we just started talking about the ships and all that and how, because the interior of the ship has a different gravitational thing [happening]. That's why it doesn't affect biological bodies.

And that's what's getting a lot of people in government [interested] right now. Different governments around the world [are interested in] how these entities are inside. But these ships are going super-fast because it has a different type of time and a different atmosphere. It's a biological type of vehicle. Yeah, it's weird to explain, but I think in GSIC [Galactic Spiritual Informers Connection Conference] Elena Danaan really explained it well. So that's quite interesting about these ships. And how people don't die inside these ships [due to high acceleration maneuvers].

MS - That was her presentation in 2022, where she talked about the propulsion system and, how gravity works, and how the laws of inertia don't apply in such situations.

Elena Danaan's 2022 presentation at GSIC was titled: "Technologies of the Star Nations," and she shared details of the different propulsion, energy, and navigation systems used by extraterrestrial civilizations.[50]

JP - And something really interesting that I think she talked about, Elena Danaan, about the [failure of] cameras, right? I could share my testimony on that. The cameras do not work around these ships because of the gravitational or electrical systems and the type of radiation that these ships give out. You know, if you go to Japan, right, and then you go to a nuclear plant, and you go into the box where the nuclear power is, and you turn on your camera, it's going to be grainy, grainy, grainy. Anywhere there's a type of radiation, your camera, your electronics, they fry, or they start graining up, and it starts not working. ... These electronics that we have right now they do not work, not unless it's from a certain distance.

When they reach a certain distance, that's when we can take pictures or take videos, [but] not even the videos work sometimes from a distance. I remember, I think you're talking in an interview that we need those cameras that they used to use in the 1940s in order to capture these ships and all that, right?

MS - ... The early cameras, they were mechanically driven. Like the operator would have to turn a crank to generate the power.... Because modern cameras or cameras with electronics just don't work. And I remember you telling me about that incident in 2018 in Orlando where the flying saucer craft landed, and the Nordic came out, and you talked, and I think, as I recall, you said you tried to take a photo or something, but your camera stopped working, but only when it left, and it was in the sky. Then your camera started working, and then you took photos, and we actually have those.

JP - Yeah, that's the one that shows a reflection of the trees in the bottom, right? Yeah. It starts working when they start going out. But I tell you now that certain kinds of these ships have a gap, right? So, it's like a bubble around them, right? An invisible bubble that does not let electronics work. But certain ships have, like, a gap that if they're in a certain position, the camera would work, and then when. They turn, that [electromagnetic] bubble will cover the location of [where] your camera is, and then you can't record anymore. You can record certain experiences on and off, and then your cameras could work in certain parts of the ship inside because they have these certain gaps. They do use certain types of technology that we have shared with them. We have their technology, and they really are intrigued. They're really interested in cameras and videos and how we can capture time on a device. So that's a technology that only humans develop, capturing people [on film] ..., but we learn how to capture time in a device and share this information.

JP - So that's quite a good technology that we have developed. We learn how to capture light and capture time. There are other advanced civilizations out there that haven't even gotten to that yet, but they have got stuff that is more advanced than us that we like to share. A lot of people and ETs are interested in capturing time and pictures and how simple it was to develop and how we developed it.... So, yeah, back to the mission, back to the experience. We all got into the van, and we started. I'm sorry, I'm talking a lot of other stuff, but it's just all mingled in with each other. Yeah.

NORDIC ASSIMILATION FACILITY

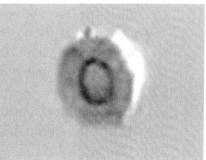

This is the fifth in a sequence of six photos taken by JP on May 24, 2018. The image on the left is the original and the upper right portion shows a close up of the flying saucer craft as it was departing, after having earlier been on the ground where JP met its Nordic extraterrestrial occupant.

Figure 18. Departing spacecraft carrying Nordic JP in USAF uniform

JP - So, we got into the van, and we started going to a facility, right? And we parked in front of this facility, and the four Nordics came out of the van. They went into this facility, and we were waiting outside, and we were talking to the driver. We're like, "Hey, are we done? Are we leaving?" And we just started talking in the van, waiting for these four Nordics to come out. I don't know why they went to the facility, but we were not allowed to go into the facility, and we were just waiting. I was on my phone. There was no internet access, so I was, you know, just chilling. I even called my wife. I was talking to Dan about different experiences that he had. He said he has been to the arks, and he has been with me, but I haven't noticed him because he wears a protective mask ...

MS - To the arks, are we talking about the Atlantic Ark and the Moon Ark, or just one of them?

JP - The Atlantic Ark. He went with me on a couple of missions, but I didn't notice him because he was wearing a mask, protecting ... his identity. Some people don't feel safe to show their identity. There are a lot of people who already know me on the ships. A couple of pilots know me. I met up with a couple of pilots before. I think I told you that we have talked about certain types of missions. I have met up with even a family member that I didn't know was part of anything, and he did a couple of missions. So, there are people who already know me about the whole mission type deal thing. So it's quite interesting.

The family member JP is referring to is named Alex, and they met during a mission to a giant spaceport and Inner Earth civilization under a South Atlantic Island on May 26, 2023.[51] After the mission's completion, JP privately contacted Alex to verify what had happened. To JP's great surprise, Alex had no recollection of the mission or meeting up with JP.

JP - So we were waiting for the Nordics to come out. We were there, like, for 40 minutes, maybe an hour, waiting for the Nordics to come out. Then four people came out, and they came in the van, and then we looked back, Dan and me. We're, like, sitting in the front. We're like, who are you guys? So, the guy that came in first, he's like, it's us. And then we looked at them again, and they [physically] looked just like us. They were not pale. Their hair was changed. They were dressed like [normal] humans. They had jeans, khakis, and boots, and they looked like us.

I'm like, "What the hell is going on here?" So, they all started talking about that Greek-sounding language. It's like, "Hey, are you guys the same people we picked up?" Yeah, and we kept looking at them. And then we look at their eyes, and they still have their blue eyes. But they look more like us. It wasn't makeup.

MS - What about their hair? Was it cut short or long?

JP - Yeah, their hair was cut short. They looked like us, Doc.

MS - Before they went in, did they have long hair?

JP - They had long hair. They looked like Nordics. They went into this facility, and they came out looking like us. Like, you would not notice [that] these people are ETs. And we're like, "Wow!" But I noticed one looked a little bit oriental, and they all had papers. And they all had passports, and we looked at them, and then the guy driving, says "All right, we're heading to the airport." I'm like, "what?" [The driver replied] "Yeah, we're going to drop them off at the airport." I'm like, "All right, let's drop them off." So, we went back, and these guys they're here to do certain [missions] ... There's one that had a plane ticket to go to Japan and he looked oriental.

So, whatever this facility is, it's a type of facility that can turn these Nordics into [ordinary-looking] humans. I don't know if it's a surgery, or I don't know if they get into [something]. They started talking about [how] they get into this capsule, and this capsule is, like, see-through, and the capsule, they lay back, and then it does everything for them. It changes them, by the way they want to be by technology. Like, Dan was, like, next to me saying, "what technology is this, man? This is crazy. They look just like us." It was quite interesting that we saw them change [to look] like us, but they looked like if you drop them off in the mall, you would never, ever realize that they came from a ship or they came from somewhere else.

MS - How tall were they?

JP - Oh, they were my height, 6' [183 cm]. One was 6' 2" [188 cm]. Another one was, like, a little bit shorter. Maybe 5' 11" [180 cm]. But they're all, like, in between 5' 11" and 6' 2".

MS - They would fit in perfectly in a normal crowd.

JP - Oh, yeah, perfectly. And we dropped them off at the airport, and their demeanor just changed. Their face turned to [a normal] human [expression], like happy, "Hey. okay, bye, guys." Like they knew us for a long time, and me and Dan just looked at them like, "Bye. Have a nice trip." And Dan just got pissed off when we were dropping them off at the airport. Like, "This is bull. Why are they doing this? Why can't they just be themselves?" I'm like, "Dan, they're here to do a certain mission, and we know about it because we were sent here to pick them up and to take them to this facility and to do whatever change they need to do, but they need to do a mission."

And I'm sure, Doc, that they're everywhere. When I saw this, my whole brain went 'boom'. My whole brain blew off different kinds of information and how everything was clicking together. There could be people in different governments. It could be doctors. There could be lawyers. There could be... Well, I know a lot of people have been saying this for a while. They walk among us, all that. A couple of drip-drop disclosures here and there about it, but when you see it, it's crazy when you see it, and it changes how you think about different things, and who can you trust? I'm not saying I don't trust the Nordics, but they could be anywhere.

This was an experience that really, I think, changed me a little bit about seeing things differently. I knew that this was happening, but when you experience it, and you see the same people change and how well they change to regular humans, you'll be surprised. You'll be quite shocked at how good they look. And it's not makeup. I even touched their skin, and it's not makeup. It's skin. They look like us. Their nails, their eyes.

So, we went back and dropped them off. We went back, and we got dropped off, and then we went to the base, and then that's when I saw. Two ships fly, like, really fast. And the last one that was at the back went really fast, and I took out my camera, and I took a picture of it, and that's when I sent it to you. That same particular ship that landed

MS - Again, that was the ship that had carried the two Nordics from the Moon.

JP - Yeah, I don't think it was the same ship, but it was the [same] type of ship that carried the two Nordics from the Moon or from wherever they came from.

MS - Okay, the photos you sent me are of a ship similar to the one that dropped off those two Nordics from the Moon. [see figure 17]

JP - Yes.

MS - Okay. And the two Nordics that came from the [Atlantic Space Ark], and they spoke, kind of like, colloquial American.

JP - Yeah.

MS - Can you say anything more about them? How can they come from the space ark? I mean, that's just more of a transit point. Right? They don't live on the space ark?

JP - No. I think they came to bring information from the space arks to other places to share information. But they had to go to this facility first by our base.

MS - Okay. Ultimately, your mission was to accompany these four Nordics. Two came from the [Atlantic] Space Ark, two from the Moon, [and you] accompany them to this facility at a base. This is the same base you took me to?

JP - Yes.

NORDIC ASSIMILATION FACILITY

MS - Okay, so this is a base you took me on a tour of last year [March 2023]. You took them there to a facility. They went in [with] long hair, fair skin. They come out with short hair. Their skin is, I guess, more tanned, and they're wearing normal clothes, and they just perfectly fit in. And you drop them off at the airport.

JP - Yeah.

MS - Did they say anything about where they were going or what they were going to be doing? I mean, how do you know that they were going ...?

JP - We were talking about how they changed [their appearance] so good. So that's when they were talking about, "Hey, they put us in this thing, and it changes us." But they never talked about their mission, not about what they were doing.

MS - ... How do you know their mission was or part of their mission was to tell other people about the space arks?

JP - Their mission, when they changed [appearance] is to share information with other people.... The way they talked was so knowledgeable of what they go through. You feel it, Doc, that they were here to bring information. They were not telepathically talking with each other. They were talking out loud about what they were doing and what they were going to do.

MS - So when you say they're here to bring information or share information with people from other parts of the world, presumably scientists or whoever, what kind of information are we talking about? We're talking general information about advanced physics propulsion systems.

JP - It is a mixture of everything. Soon, the public will know about certain things. And it's just certain things that they're talking about, I think. I can't say now. But, yeah, there are certain things that they were talking about, and we knew that they were visiting other places to bring information.

MS - That's very interesting because I know there have been other cases, contactee cases from the 1960s, where you have kind of Nordic-looking extraterrestrials setting up a base of operations where they blend in. They're helped to blend in, and they go and share information. That was their job, to just find people and share the information to help raise the planetary information level, if you like.

JP - Quite interesting. And that happened, what, in the year? The 50s, the 60s?

MS - There are multiple cases. [There was] the Howard Menger case, where he had something similar [happen] where he was helping Nordics prepare to assimilate into American society. He would give them haircuts, and he would help them get clothes and get identities and all of these things to fit in. And there's another case from Italy in the 1950s called the Amicizia case. Again, similar thing. So sounds as though this continues to happen. This has been going on since the 1950s.

NORDIC ASSIMILATION FACILITY

In his 1959 book *From Outer Space*, Howard Menger wrote about the assistance he gave to human-looking extraterrestrials to fit into normal American society:

> I remember several occasions when I cut their hair. I don't know if they saved their hair or not. However, all evidence of the meetings was always carefully gathered up by the space people before they departed. The men, particularly the Venusians, had unusually fair skin, without hair on their arms or face, and had no need to shave. After three months on Earth, however, they became hairy and grew beards. Most of them waited the three-month period so they could have beards to shave and appear more like earth people. Some of them requested dark glasses, a few asked for dark glasses with red glass, which was difficult to obtain. I don't know why they wanted dark glasses for those I had met previously had not worn them...
>
> I briefed them on our customs, slang, and habits. Although they utilized instruments to learn a language quickly the machines couldn't always cope with colloquialisms. And they had to pass for ordinary people. I was gratefully surprised to be of immediate help to these advanced people.
>
> ... They never asked me to obtain any identification papers for them or to help them locate jobs. They seemed to be able to take care of such matters themselves after they had been properly acclimated and grown accustomed to our ways. Once clothed in our attire and briefed thoroughly in our customs, they were on their own, and seemed to experience no difficulties.[52]

Similarly, in Italy, from 1956 to 1978, human-looking extraterrestrials interacted with up to 200 prominent Europeans from Italy, Germany, Switzerland, etc., and several helped them to assimilate into normal human society[53] The extraterrestrials were based in underground facilities near the city of Pescara, Italy, and other areas adjoining the Adriatic Sea. Once again, the extraterrestrials received much valuable assistance on how to fit into normal society. Importantly, some of the individuals helping the extraterrestrials were former or current senior officials in the Italian or foreign governments and security services. This meant that NATO was closely monitoring the situation and learning about the processes used by extraterrestrials to assimilate into human society. This official monitoring process culminated in government and military agencies developing their own facilities for assimilating extraterrestrial visitors into human society. The former Prime Minister and President of Russia, Dmitry Medvedev, spoke of such extraterrestrial monitoring and assimilation facilities in a startling hot-mike incident in 2012.[54] Now, 12 years later, JP had witnessed one of these worldwide extraterrestrial assimilation facilities at a major US military base.

> JP - Oh, yeh, I believe, by the way they did it, and by the way, how normal they were about it [after leaving the facility], it looks like this happens all the time. It's just my first time experiencing it, looking at it firsthand. I think they wanted me to know that this happened just to bring it to the public's [attention]. "Be careful who you talk to. Be careful. We're around, so we don't want no negative people doing negative things." They're around, I guess. "Be careful you don't do bad things." I guess they're all around [watching]. So, it's quite interesting.
>
> MS - Now, did you get a chance to ask them more specifically who they are? I mean, people want to know, when you describe them as Nordics, that's

just the physical appearance. Scandinavian blue eyes, hair, fair skin, look. And, you know, two of them came from the Moon, and two came from the Atlantic Space Ark. So one of the questions is, are these Nordics extraterrestrials from ... other worlds? Are they from the [Inner] Earth? People would like to know, are they Pleiadians, are they Sirians, are they Andromedans, that sort of thing?

JP - It's more like we always had a feel, like from the Pleiades, that feeling.

MS – So, you just have the feeling they're from the Pleiades?

JP - Yeah, I do.

MS - The ones that you've been in contact with since 2008, do you have a feeling they're from the Pleiades, or did they just say where they're from?

JP - No, they never said where they're from. But it's just the way people describe how Pleadians are, they act similarly to it. So. Really, Doc, I don't know where they're from, but I don't think they're from the Inner Earth. I think they had integrated into the Inner Earth or onto our planet and mingled around with us, yeah, for thousands of years. But do not think that they're from planet Earth. I think, yes, they do come from somewhere else.

MS - You said two of them came from the Atlantic Space Ark. Do you want to share anything about what they said?

JP - No, they just discovered something in the Space Ark that they want to share with somebody. So, I know we took them to the airport, and I know one was going to Japan. That's the only location I know. But the other ones, they're probably staying in the States or going somewhere else, I don't know. But there's one that turned oriental looking. And he was talking about Japan. So, I understood that he was going to Japan. They're all going to different locations. They're all connected with each other. They have devices that connect them with each other. It's quite interesting.

MS - Okay, so any final things you want to say about this particular mission?

JP - It was [life] changing to see this. I think for some people, it will be nerve-wracking to see a situation like this, but I think they're here to help. Doc, I don't think they're here to confuse. Well, yeah, it could confuse people when you see them turn into human beings. And all that, but they can't be themselves yet. They can't show themselves yet. There will be a time that, yeah, they're going to be walking among us, and we're going to see these people walking around us with technology that we never knew existed.

MS - Can you explain how you were tasked for this mission? I mean, did you get a phone call? Did someone tell you to physically be at a certain location to do this and do that? And was there any debriefing?

JP - There was no debriefing. There was nothing like that. About the [mission], it was a type of task. And, yeah, it was just a task. You go to the base, and [it's like], "Hey, let's go! You get into the van, be at this location at this time!" And we got Into the van, and we went to this location, this field, and the rest happened, you know.

MS - I don't know if it's okay for you to maybe explain how the process happens. I mean, you got your normal, regular army duties, but does someone come up to you and say, "Hey, JP, go here and be at this place." And they're part of a covert branch, covert missions. Do you want to elaborate?

JP – Yeah, you get a message through your phone?

MS - So it's a text message?

JP - Yeah.

MS - I see, okay. So, you recognize that when you get a text message from a particular phone number, this is part of the covert mission.

JP - No, it doesn't say covert missions. It doesn't say anything like that. It just says a certain number, a certain code that we know.

MS - I see.

JP - And that's it. It's only a number. Okay.

MS - And so after the mission, you get the green light or the red light from someone who's a more senior official in this program.

JP - It's because I'm bothered [by the mission]. Sometimes, I don't like to keep stuff like that to myself. So, I will go to the location where I took you [in 2023], and I will meet up with this particular guy. He's like, yeah, you can talk about it, but don't say certain things. Just say this and that. He's like a coach type of guy and tells me what to say or what not to say.

MS – Okay, so we know he's an officer. He's kind of like, we can describe him as your handler, someone who makes [sure] that you stay within the parameters that they set for you, being able to reveal details of your missions, but not revealing too much, that could get you in trouble or expose too much about the covert leadership.

JP - Yeah. I know a lot of people will know what I'm talking about that hear this program and hear this interview. They'll know what I'm talking about when I say that you can't say certain things because of national security reasons, but there are some things that we can't just say full-blown out. Not unless somebody higher tells you that you can say, but, yeah, I think I said enough. Right? I think that's a lot of information. I think that a lot of people did not know that these ETs looked just like us.

MS - Well, yeah, that's very fascinating that there's now an official program to help these Nordics or these human-looking extraterrestrials to assimilate into human society. I know of cases, as we talked about … where contactees and private individuals were helping the Nordics assimilate. But now we know that there's an actual, official assimilation facility at that base that you took me to back in 2023.

JP - And I'm sure, like in the GSIC, there were a couple of them walking around.

MS - GSIC stands for the Galactic Spiritual Informers Connection that was held in Orlando in October. And you were there secretly. And there's going to be another one in 2024. ... That's going to be in Denver, Colorado. I guess we can't say anything about that at this moment. Yeah, but some of these beings were there at GSIC, and that's what you're saying.

JP - Yeah, there's a couple of different characters that we met there. That was quite weird. Yeah.

MS - Okay, well, I guess there's been some development. Now, you have an Instagram account that we are announcing that people where you can post directly information that you find pertinent. So, do you want to talk a little bit about the Instagram account you created and why you created it?

JP - Sure. To connect with people. After this beautiful interview, people can connect with me over there on Instagram. And it is me. A lot of people are asking, "Oh, is this the real JP, or is it somebody else?" No, but yeah, it is the real me—JP.missions. I'm there showing love to everybody because that's what we need to show right now—love and share positivity in everything that we do.

And I think the time has come for me to do that. For me to connect with beautiful people and people who have been through similar things that I have been through. So, yeah, I think this will open up doors for other people who have been going through the same thing I have been going through because we're all, sooner or later, going to be coming out all at once. But yeah, I'm there on Instagram. I'm also doing a YouTube account, but I'm not going to post any videos there, ... [until] more in the future when I use a YouTube account.[55] But the Instagram account I'm using right now. And, yeah, I'm there for any questions or anything that you need to know about the interviews that we had. You can ask me there, and I'll briefly answer you back.

MS - Okay, great. Well, that's very encouraging. So any final words you want to say to our Portuguese-speaking or Spanish-speaking friends?

JP - ... So, yeah, basically what I said is. On the Instagram page, I have a link connecting to you, Doc, about my experiences so. Yeah, it has the books as well that you wrote about my experiences. Beautifully done. And right there, they can clearly see the proof and the pictures and the videos and all that since 2008, my experiences. So, yeah, I'm really happy I brought this information out, Doc. And I appreciate you, Doc, for all the work that you do for humanity. This is special that we're doing what we're doing, and you're chosen for the right time. All this is coming out because I know a lot of people who have been going through this about ETs and about different civilizations, and different experiences and have been going through tough times. But now it looks like the veil has opened, and everybody's talking about it.

So, all those people that thought that we were nuts, I guess now they understand that this is real, that this is out there, that there is something going on, and that sooner or later, everybody's going to know about it. In Brazil, in different countries, and in South America. Like, everybody is recognizing [this in] Europe, that this is happening. So, yeah, that's my message to people. And just love and kisses and hugs.

MS - A very important message. Thank you, JP, for your service and for sharing information from your missions. It's a privilege to know you and to be working with you. So, thank you, and I look forward to hearing about future missions.

JP - Roger that. JP out.

JP's report that Nordic-looking extraterrestrials are being assimilated into Earth society through secret facilities found in major military bases is nothing short of astounding. It confirms what many others have previously said about extraterrestrials living and walking among us, as documented in my 2013 book, *Galactic Diplomacy*.[56] In it, I cited multiple sources revealing extraterrestrials have been infiltrating human society since at least the 1950s with the help of private individuals. This informal assimilation has likely been happening far back into human history. From what JP has revealed, we now know that such informal assimilation practices have been taken over by select government and military agencies who have developed sophisticated tools and institutions for assisting and tracking extraterrestrials living or working among us.

Chapter 7

The Ancient Underground Castle with Gold Plates

On February 20, 2024, JP was sent on a five-man mission to an underground location in an icy cold region of the US, which we later calculated to be somewhere in North or South Dakota. He was taken by a Blackhawk helicopter, where he and his companions were drugged and woken up when they arrived at the destination. They were dressed in black outfits equipped with radio-connected masks and proceeded to enter a giant underground cavern that was very cold with ice. They saw a large underground castle-like building that was covered by ornate gothic-style architecture with many mythological figures depicted, including gargoyles.

There were other military teams at the location filling shipping containers with gold plates, each weighing around 25 lbs (11 kgs) and shaped like a frisbee. Each shipping container was marked with a different destination around the planet, suggesting that a deal had been reached where the gold and artifacts from the ancient castle would be shared with an international alliance of some kind. Is what JP witnessed part of a secret Earth Alliance plan to launch gold-backed currencies to free humanity from debt-laden FIAT currencies?

JP next witnessed several very tall beings, over 6 feet 6 inches (2 meters), also wearing masks and guarding a doorway, who he suspected were non-human. JP talked with a short-haired Nordic who said he was there to study and catalog the artifacts. There was also a Gray-like entity that was studying the artifacts and watching the actions of the different teams. JP's team proceeded to load one of the shipping containers with the gold plates and finished when it was loaded.

JP supplied two illustrations of his military team traveling to the underground location and the castle, and he took a video of the end of the mission when he and his team arrived back on helicopters at night. He concluded that his team's role was to observe and reveal the mission to the general public. The transcript of the interview follows, with grammatical corrections and my commentary.[57]

Interview Transcript with Commentary

Key: MS – Michael Salla; JP – Pseudonym for US Army Soldier

> MS - It is Thursday, February 22 [2024], and welcome JP to Exopolitics Today.
>
> JP - How are you doing, Doc? It's a pleasure to be here on Exopolitics [Today] to bring this beautiful information to you.
>
> MS - Well, I'm very glad that you're able to give us an update. It's been a few weeks now. So yeah, why don't you tell us? What is it that recently happened?
>
> JP - Well, I just also want to thank the public, you know, for following me and being there for me on YouTube and Instagram. It's really beautiful. The people who are starting to follow me are from Brazil, all over Europe, the United States, and South America. It's amazing the support, and I just want to thank the public for that. We were heading to a field, right? Everybody got into like an SUV, a total of like [seven] people, three people sitting in the back, two in the middle, and the other two, the one that was driving and the other one that was sitting

THE ANCIENT UNDERGROUND CASTLE WITH GOLD PLATES

in the [front] passenger seat. We took an SUV, and we drove more into a practice zone.

So, we drove there, and we had all blacks [black outfits]. I knew something was gonna go down. You know, when you get that feeling, "Oh, okay, something is gonna go down." So, we parked, we got out, and then, out of nowhere, four beautiful helicopters, patchy, Black Hawk helicopters started landing. And I'm like, man, what's gonna happen? Dan was there with us, but he had a mask on.

I didn't see his face, but I knew his voice. I said, "Dan, is that you?" He's like, "What are you talking about?" I'm like, "Dan, is that you?" "Yeah, man, shut up. Yeah, that's me." I knew his voice. So, he was there with us on this particular mission that we were going to.

MS - JP, do you ever have to wear a mask on any of these missions? Were you wearing a mask, or is it just some of the soldiers?

JP - It is voluntary, but they do want you to wear the mask. It's not one of those COVID masks; it's a mask that's connected to your ACH.

MS - Okay, you want to explain a little more what that would look like?

JP - A skiing mask similar to that, and it has like a little round filtration system in it on the right side of it, and it gives space to your mouth, like one or two inches so ... people will still understand you, but these particular masks are connected [by radio] to everybody, so everybody can hear you clear like this, but if you're not wearing a mask you'll hear people like [differently]. So, it's kind of interesting how everybody's connected. [Each mask] has a mike to it, but it's not made of like a fabric; it's more like a plastic type of mask

MS - Okay, so you and the other five guys were all wearing the masks?

JP - You could take it off. It buttons onto the particular helmet that we wear. Yeah, it's interesting what they give us when we go out to these particular missions because I think they know the type of elements that we're going to run into. We started stepping into the helicopter, and the driver said, "Oh, these helicopters are here for you guys." And we're like, "Oh, shit, oh, man, oh, crap. All right, we didn't know about this." I said, "Yeah, you're never going to know about this." We went where the helicopter landed, and we started going. You know, everybody ducks their heads because those helicopters, you know, those propellers, you need to be careful.

So, it's an instinct that you have to lower your head and not to get [hit when going] into these helicopters. So, we got into these helicopters, strapped in, and waited like 10 minutes. There was a medic there giving us shots. I know a lot of people are probably going to be mad about this, but I assure you that we do everything and take precautions in everything that we do. And you trust each other. That's the main thing about being in the military: you have to trust each other. And if the medic wants to give you a shot for this particular mission that you're going out to, you're going to take the shot. You know, so I took I took the shot.

MS - And there are four Black Hawk helicopters, you said, but all of you got onto one.

JP - Yeah, there are already other people. You can see other people already there. They had picked up other people from different spots, but I noticed everybody seemed not dazed but sleepy in the other helicopters. So, I'm like, "Oh shit, this is going to put us down. Whatever this shot is". So, I got the shot, and I felt this taste under my tongue. It was a weird, like a vinegar taste. And I just, we just, dosed off. I don't know what time we were on. I was already on the helicopter strapped on, and they were giving shots to everybody. And I guess after 10 minutes, I just blacked out. And I don't remember anything about how long the trip was or how long we were at this place.

When I opened my eyes, they cracked a "waker", and they put it to our nose. And I just woke up. My eyes were just wide open, and everything was white. And I heard the helicopters; the other two helicopters had taken off, and you could see the other people. There were Marines, there were Air Force personnel. There were people from different countries dressed in different uniforms. We were in this big, ice-looking cavern. I sent you the pictures on WhatsApp with an idea of how it looked.

MS - So those pictures you sent resemble how it looked, but they weren't [photos] of the actual incident?

JP - Just resembling how it went and how it went down. So, we saw this big cavern and another helicopter landing in the middle of the cavern. It was cold as shit. It was really cold. I'm not saying it's Antarctica, the North Pole, or the South Pole, but I know it was a cold place. I know North America is in winter right now, so it could have been anywhere in Canada. That's the closest thing I can think of where maybe this type of cavern would be, and everything is frozen.

MS - Now, just to understand this, I mean, you got on this Black Hawk helicopter with five other guys. There's a bunch of other Black Hawks; you get the shot, then you wake up, and you're in the Black Hawk. This other location, I mean, did the Black Hawk do that? Because the Black Hawks can't go that fast, I mean, to go from...

JP - It must have been probably a place that was really close. I don't know, I blacked out. I don't know how much time went by. I was looking at my watch, and it was showing at the same time as when I left. So, the situation that we were in didn't make sense to me. So, it could have been that they took us to another location, put us on a TR 3B, and everybody, I guess, was asleep. They then took us to another location and then put us back on an Apache (Black Hawk). But that would be too much work, so I don't think that was the case. I don't know where it was. I can't say that it's Antarctica. I don't know where it was, but I know it was a cavern, and it had it had ice on it. And you can see a waterfall that was frozen coming in from the hole where the helicopters were coming into and landing. And it was a cold place.

THE ANCIENT UNDERGROUND CASTLE WITH GOLD PLATES

Figure 19. JP illustrated the scene he witnessed when arriving at the ice cavern

JP - And it puzzled me the same way as puzzling you about where the location might be because a helicopter cannot travel that far, that fast, or I don't know how long it took. Yeah, they don't travel that fast. So, it must have been somewhere close to Florida, but I don't know where it would be frozen like that near Florida or near the lower states. Not unless maybe a mountain that's located with snow in the Georgias or Carolinas.

After JP's shared his mission report with me, we calculated that based on the total time he was away on the mission and the Black Hawk's maximum speed, that he was likely taken to some location in the Dakotas.

MS - Oh, some sort of icy cavern somewhere where there's snow at the moment?

JP - I do not know where the location was, and my phone was not working, and my watch was not working. We were confused, like, "Where the hell is this place?" I remember we were in a high, snowy place, but we were inside a cavern. It was really nice. You could see the sun peeking into the hole, and there's like a river, and then we walked more into the cavern, like probably three football fields into the cavern, and then we see this structure that looked like a temple, but it looked beautiful ... with so many details, but you could tell it was like a type of granite marble, made of granite marble similar to how they build in Spain. I don't know if you're familiar with Sagrada Familia Church over there or the type of detail they have over there.

MS - The Gothic-type style of architecture, yeah.

JP - That has a lot of detail and all that, but it was embedded into the cavern. It looked like a little castle. It was beautiful, and there were crates right next to it. And they were putting stuff into them. There were a lot of different soldiers and a lot of different people putting these boxes into it. They called us down, and we started going down closer to this beautiful building or castle or whatever you want to call it. It had a lot of windows. It looked like it was vacant; nobody was living there when we entered. It was probably like 14 or 16 feet high, the arch when you enter into it, and then it grows into like a 20 feet [6.1 meters] high arch. When you enter the building and you look around, it's beautiful. A lot of statues of mythological creatures ... Greek type of mythological creatures. A lot of the gods of the Greeks. Greek gods, I saw a couple of them, and then it ... transitions to the Mayan type of dragon-looking gargoyle to the sides. It was a beautiful, beautiful, beautiful place. It looked like a type of history, a type of place that people would go to, like a type of museum of statues.

MS - So this kind of castle-like building is inside this massive icy cavern.

JP - Yeah, it's inside this massive icy cavern, but everything is frozen. I think because there's water going in and the temperature gets lower inside the way it is, it gets really low in temperature. Everything freezes connecting to the structure, so it was cold. I can tell you that it was really cold. You can feel the wind still in the cavern, but you can [also] see the sun. You can still see the whole of the sun coming in. You can see the reflection of the granite and all that. So we went in, right? And you can see a lot of soldiers working in different spots of this [cavern]. I'm like, "Man, what is this?" I was freezing. I'm like, "What are we doing here?" And then it's like, "Bro, shut up and just go with the flow, man. Just go with the flow." I'm like, "This is too beautiful to be kept like this, you know?" I think people should know about these particular places.

Figure 20. Sagrada Familia, Barcelona. Source: Adobe Stock

JP - It's like, sometimes these places cannot be disclosed because of national security issues. I'm like, "Oh, okay." And the security of those that built it, I guess whatever civilization built this castle tower type of thing. In the middle of the structure, there's a type of stairway that looks like a tunnel that goes in really deep, but it has a round door on it. You can see a metallic round door that was blocked. And we saw at least 30 soldier-type people. And they were bigger people than us, probably 6' 6". They are protecting this huge door so nobody can get close to this door, but we can still be close to the structure.

When I saw the crates of what the other soldiers were carrying out, there were round gold plates stacked up. Gold, a lot of gold, a lot of gold. I [have] never seen so much gold like this. I'm like, "Man, if this is ours, this country just became the richest country on the whole freaking planet. If all this gold is ours." The room that I saw had all these plates of gold. [There] were probably thousands, hundreds of thousands of these plates of gold. And people were taking them out, putting them in crates, and putting them in other crates, bigger crates. Like, you know, the one people use when they put on boats to take to other countries.

MS - Pallets. You're talking about pallets?

JP - No, that's not pallets. They're putting them in massive crates.

MS - Like the containers?

JP - Like the containers. Yeah. They have at least 30 or 60 containers all set up like this, and you had types of trucks that were coming in and heading out through another location that we didn't know where they were going.

MS - So when you're saying those containers, we're talking like those big 40-foot-long shipping containers? So, they're being filled with all of these gold plates?

JP - Yeah, a lot of gold plates.

MS - Can you describe what one of these gold plates looks like? I mean, how big?

JP - They're the size of a regular frisbee, but they were solid. They were round, the size of a sports frisbee. But they're thick and round, and they have weird ancient Sumerian writing. Lines with triangles on each one. So, Dan, he was telling me, "Hey man, I think this is a type of freaking payment or something." I'm like, "Bro, but payment from where?" And he was like, "Man, I heard something about them striking a deal or something like that with technology." He was just talking some random, weird shit that we struck a deal with an Inner Earth civilization, and they told us about this location and that we have to, I guess, split the profit with everybody else. So, I'm like, "What the heck is going on? Like gold and only gold. That's it." And we were like, "Bro, this is a lot of freaking gold, like a lot of gold." And it was that dark, nice gold. Like, you know, that it was real, real gold. I picked one up, and it felt like probably 25 pounds [11.3 kgs].

MS – Okay, so just one of those gold frisbee-shaped plates was 25 pounds. Okay, so I guess that would be like solid gold.

JP - Yeah, so we helped a little bit with the crates and putting them in the container and each container had a different language on it. I remember seeing an Asian type [of writing]; I don't know if it was Japanese or Chinese. But it looked like an Asian [language] on the crate. I saw something that resembled Greek on the crates, also Arabic on the crates, and even Hebrew [was] on the crates. So I think these crates were going out to these different countries. That's me assuming, you know, "Why else would the name be on the crates, on the containers and all that? So, we went more in, and I saw Nordics in the distance, and they didn't have long hair.

MS - Before you talk about the Nordics that you see in the distance, you first described some guards at some point in that castle. There were some guards, and you said they were different from the soldiers [and] your guys there.

JP - They were huge guards, and I did not want to approach them.

MS - Was there anything distinctive about them? Were they human? Were they something else?

JP - I don't know. They were wearing uniforms, and they were covering their faces, but they looked huge and chunky-looking, like big people. But they were all covered up.

MS - But they were different to the soldiers? I mean, the soldiers that you're with and those who were there [before your arrival]?

JP - And even the weapons were different types of weapons that they have. It looked like a gun, but the opening of the gun looked like a flower-looking thing that closes up in the tip of their weapon. They had something that closed up like a flower. Like a rose before opening up, you know how a rose looks … They told us not to get close to them. So, we decided, "Yeah, let's leave those people alone.

Whoever they are, we don't want any conflict with them." So, we just did not want to approach anybody who had weapons and protecting the door. Let's leave them as they are. You know, sometimes when you go on these missions, they tell you not to approach something. Just don't approach them.

MS. - So these beings, they were guarding some doorway to some other area of that castle-like structure.

JP - And you can see stairs, and then there's a big metallic door, like protecting [something]. So, on the stairs, and you have like that, there are at least 20 of them protecting the perimeter, and they're just walking back and forth, talking to each other with a really low tone of voice.

JP - I heard English, but there were also different types of language that they were speaking. I don't know what it was. So, their uniform was black as well. And they had a type of metallic thing coming out from the right shoulder, and it had a beam that I think was a type of light thing, magic that they could turn on.

JP -So we saw that, but it was not turned on. Yeah. They looked different. Yeah, I [have] never seen them before. I don't know who they are or what they are, but I did see the Nordics. They were in a military uniform, their hair was cut off, and I wanted to approach them. They were talking to other soldiers, like nothing [was strange], telling them what to do and how to do it. They were also there to investigate the statues and the language in that building. So, they were doing research on that, too.

THE ANCIENT UNDERGROUND CASTLE WITH GOLD PLATES

Everybody was working together. I did see, but I don't know if I should say Grays.

MS - Before you talk about the Grays, let's just back up a little bit. You say you saw the Nordics there as well.

JP - We stayed and talked a little bit, and all that, about the situation that was happening around the world and how they were involved. We just asked random questions and said, "Hey, man, how do you guys fly around those ships?" And they started laughing. "What ships are you guys talking about?' 'Yeah, those ships that we see flying around, beautiful, beautiful spaceships that you guys fly around." I don't know; this is like an inside joke. When you see a Nordic, you always start talking about the ships. So, they get annoyed sometimes because they don't like talking about the ships. I went close to the one and said, "Hey, can you tell me any information about these ships? He's like, ha, ha, ha. What do you want to know?"

I was like, "Hey, how is it inside?" So, he looked at me really seriously. He's like, "What do you want to know?' I'm like, "No, we just, you know, we see you guys flying around a lot, and I really want to know, about these ships, about how they fly?" He's like, "Yeah, but what do you want to know?" Like, "Do you guys have gravity inside the ship?" And then he told me that each planet has a distinct gravitational pull, right? And each planet has a different oxygen density. So, the ship mimics the planet's gravitational pull inside the ship.... He was telling me that the ship mimics the gravitational pull of the vibration of each planet. So, it gets embedded in the

DNA of the ship. Say a ship enters Earth's atmosphere, right? The ship still mimics the planet that they last visited. The atmosphere and the gravitational pull [remains] inside the ship. But if they want to change it, they have to go through a heating process. And I'm like, "What do you mean a heating process?

So, the way these ships work, they have to go through a heating process in order to mimic the same gravitational pull and oxygen density that the planet has. So, they can come to Earth and then leave Earth's atmosphere, and they still have gravity inside the ship and oxygen density inside the ship, as if they were still on Earth. So, I guess the ship has a way of mimicking and embedding the type of gravity and the type of oxygen level that the ship visits.

It can be any kind of planet. So, each planet is different in size. So, if you do see a Nordic that is probably eight foot tall or seven foot tall and weighs like 300 or maybe 350 pounds, their planet was slightly bigger than ours in the Goldilocks zone of their solar system. There are planets that look like ours, but the organisms are bigger because the planet has a bigger gravitational pull and probably has a higher oxygen density. I wasn't understanding as much, but I kind of got the idea about how maybe these ships mimic the gravity and mimic the oxygen density of each planet that they visit, and it stays ingrained in the ... [ship's] memory. So, when they do visit these planets, they don't have to go through that heating process. The ship knows what planet they're on, because I guess it matches the data from the first [time]... Are you understanding a little bit?

MS - Yeah, I get that. So, the environmental and gravitational variables for different worlds and locations are factored into that craft's navigation system. And so when they go to different places, they're prepared in the interior of their spacecraft for the exterior things. But I assume then they might have environmental controls like we do like they might have masks or something. But it sounds as though these Nordics that you were talking to were breathing the same oxygen and walking around. So they're completely acclimated to Earth conditions.

JP - Yeah, they're really acclimated to Earth's condition. Then there are different oxygen levels also in Inner Earth type of civilizations. Underwater [civilizations have] also a type of oxygen density that is different. So yeah, it could explain why these ships, when they enter the water, keep the same speed. And they keep the same gravity inside the ship and oxygen level. So yeah, it was quite interesting.

JP experienced the oxygen density levels in both Inner Earth and underwater civilizations, which he has visited during many missions. He noticed different smells and some variations in oxygen levels, but they were similar and comfortable enough for him and the personnel accompanying him not to wear oxygen masks.

MS - What were the Nordics doing there? It seems like your team was there to observe or to help in the movement of that gold into the containers. What were the Nordics doing?

JP - They were investigating the statues and the language that was there. So they were just investigating the area, I guess. They had never been there before. They never saw anything like this before. So, they were just putting one and one together trying to figure out who were these people [who built the castle] and why was everybody else receiving this type of gold, and what is this gold. This was real gold, and there's a lot of money involved here. This is the most gold I've ever seen in my life. Quite interesting this particular mission.

After we talked to the Nordics, I saw what looked like Grays in the distance, but we did not approach these entities. They were wearing all black. I could see their big eyes in the distance, but I didn't know if it was a helmet because it was too shiny. I could see their long fingers while they were talking to some other soldiers. I think they were marines that they were talking to or communicating with. They were really far away. I could see them probably like 70 yards away, but they're distinct-looking, you know. They got big heads, bigger heads, yeah, they were like pointing to the marines. Like what was happening, and I see no ships around [them] ... but I could see them talking to a marine uniform type of military personnel. I could tell that it was marine because it was a darker greenish camouflage color.

MS - So these Grays, I mean some people talk about, of course, Gray extraterrestrials, and then there are also programmed life forms, in other words, Grays, that are cloned by the space programs or the secret projects. Do you have any idea were these extraterrestrials, or could they have been programmed life forms?

JP - Do not know, I wasn't close enough to tell, but I knew they looked like Grays, but I don't know. And we were asking a lot of questions. The first conversation we had was with that Nordic because he was working with some of us and other people who were investigating the artwork and the language that were part of the military. So, we just took advantage of it, got close to them and started talking to them.

MS - So, was there a clear mission goal that you guys had? I mean, the five of you, what was your primary objective in going to this location?

JP - We were helping out. I moved like probably 18 crates of gold. It took me about an hour and 30 to move those crates around there. It was nonstop with all these guys. A lot of heavy work. It was just moving them from one place to another, making sure they go into the right place, the right crate. You have higher [level] people showing you what container they were going into. Yeah, we were just there helping out. And I guess they only use people who have been through particular missions. And they use people who are comfortable with these types of situations and missions. So I guess everybody that was there, and there was a lot of people, I can estimate there was probably like, in all 150, military personnel. But that's including the Nordics and the Grays and those other guys [standing guard to the metallic door].

So, there are 150 people in all there with other four-wheeler types of vehicles that they have, like six wheels. The situation was quite interesting, Doc. Some guy came, I think he was a higher-ranking

> dude, but he had a mask on. He's like, "Hey, we're not going to need you guys anymore. I think we got this handled." And then Dan was like, "Man, what's the freaking point of this? What the hell is going on? What was the freaking point of this?" You know, and he was just confused the crap out. He was like, "They can have other people doing this crap. You know, what is the point for us to come down to here and all that." So, when I was telling that same thing to the guy that gives me the green light to talk about these things, he said, "They want you to talk about this." That's the main reason for all these missions. They just want you [JP] to talk about this and let the public know what's going on and that this is happening, keeping a record that this is happening.

What JP was told by the senior officer green lighting to share missions is critical to understanding why he is allowed to reveal his missions. He is part of a covertly sanctioned disclosure program approved by those running secret space programs and Inner Earth missions. JP's mission reports are designed to acclimate the general public to the full scope of what's happening in Deep Space, under the Oceans, and the Inner Earth. In US Army Insider Missions 1, I explained how the Nordics and the USAF had reached an agreement concerning JP. The USAF would allow JP to participate in many classified missions and reveal these, and in exchange, the Nordics would help the USAF to find and understand advanced technologies.[58]

> JP - So when this does come out, it won't be a shock to a lot of people who are awake and know this information. Some of these missions do not make sense sometimes, but when I'm here talking to you, Doc, and we put this out and other people hear this, and it's probably other people that have been going through this too, they do feel a type of release. They

feel a type of happiness that this information is coming out because they know, you know, it's not kept from them. So, they feel a sense of release when they hear this, the people that are doing these missions, and they hear these interviews. They know that people in other nations are hearing about this information. This is not only happening over here in our homeland; it is happening in other countries as well, these types of weird missions that people go through....

We finished [loading the gold and], they told us to leave. And we were heading back to the top part of where the helicopter lands. And we waited 40 minutes up there, cold, cold, cold. [The] helicopter came in. We got on, and they gave us the shot again.... After 10 minutes, I started dozing off. We woke up again. We ended up back to where we began with the SUV. And Dan was like, "This is bullshit, man. This is crazy. Why? I don't want them to use me anymore, you know. Why do I have to go through this?" [I say] like Dan, "just run with it." He's like, "Shut up, man." So, I said the same thing to him that he told me, like, "Just run with it." He's like, "Yeah, but I just don't understand. This is senseless. Like everything that happens sometimes is for a reason."

I'm glad I talked to that Nordic. I got something out of it, you know, about probably how their ships run and all that. It was quite good that I got that these guards were different from us. And they're probably a different civilization that we don't know about that lives on Earth. And we don't know what type of technology that they have. And we don't know why [they were there]. I assume that the whole building is theirs—those people who guarded the entrance. Dan said, "I think it was like a payment of some sort to the outer [surface] Earth people, some sort of deal that went down."

Figure 21. JP's illustration of the scene when leaving the giant ice cavern

MS - Yeah, that's very interesting. I kind of wonder what kind of a deal was struck because it sounds like all of that gold that was being collected was going to be shipped off to different parts of the world. But even though that's an ancient civilization that was found, or that where the gold was found was in this ancient structure, castle, whatever, in this location somewhere in northern parts of the United States, that they would ship that to other countries. So yeah, that would suggest some kind of agreement is happening. Maybe this is being found all over the place. Similar things are happening in other countries where some of the stuff is being shipped to America. That's my guess.

JP - Yeah maybe a technology switch, I don't know. But we were puzzled when we came back, and Dan was like, "Yeah, I don't get this shit, like man, I just want to get out." I'm like, "Bro, just run with it. We came back like any other day." We saw the helicopters leaving. I took my phone out, and I took a video of the helicopters leaving the area, and I was like, "Go, go, go." Yeah, the mission was quite interesting.

MS – Okay, so you did send me a video of the helicopters. It was dark; you could hear helicopters, and I could hear someone saying, "Go, go, go." So, is that video something we can release with this recording or something we have to hold on to?

JP - Yes, it could be a release. "Let's go, let's go!"

JP's video of the departing Black Hawk helicopters was made available on YouTube on February 29, 2024. Taken during the night, one can see lights and the helicopters as the camera zooms in. One can also clearly hear helicopter rotor blades and JP shouting, "Let's go, Let's go."[59]

MS - Oh, that's great. I think people will be happy about that. So, any final things you want to say about that mission? When did it when did it happen? I mean, today's Thursday. So, we're talking two days ago—Tuesday, 22nd.

JP - Two days ago.

MS - Okay, so this happened on Tuesday, February 20. Okay, that's great to know. So, you have created a YouTube channel and an Instagram channel. You want to maybe say to the audience who aren't familiar or haven't found that yet, you know, why you created those channels, and what you plan to put up on those channels.

JP – [On] these channels, I will be putting out hints about my next missions and probably thoughts that I sometimes get about situations in life and how people should think when they experience these situations. I think it will be good. I'm connecting with a lot of people who are saying that they've been through these types of situations. By keywords, when I talk to them, I know that they've been through these certain situations. By me vetting them in a certain way, I could tell if they were telling the truth or not. So, I think a couple of people are genuinely going through these types of things, especially in Brazil and other countries. So it's quite interesting the connection.

And yeah, putting things out, and you know the real updates will always be with you, Doc. After I drop an update with you and after this information comes out, I will put other hints about things that are happening around us and that are soon to be real. So yeah, I'm really happy. Yeah, my Instagram is jp.missions, and then the YouTube video [channel] is jpjpjp1

MS - That's great …. So, any final words you want to say to our Brazilian and Spanish friends?

JP - So yeah, I basically said that I'm really happy with the support, and we're going to bring more information out just so people don't have to be in the dark, you know, about what's happening. So that's the most important thing, you know, to share this information and people will be happy about what they hear.

MS - Okay, well, fantastic. Thank you, JP, for sharing. I know people always appreciate you giving updates on what has been happening, and I'm sure people will be fascinated to learn about this really incredible discovery somewhere concerning all of this, kind of like gold that's being discovered and the agreements that are behind it and the different ET groups that are also kind of monitoring all of this. So thank you, JP.

JP - Roger that, I appreciate you, Doc, for everything that you do, and yeah, people are going to be shocked when all this information comes out to the public about these different types of things are happening worldwide and in our solar system.

To date, JP has not been told the reason why the gold was being shipped out of the US to other countries and the precise nature of the agreement between the US military, international governments, the Inner Earth Civilization, the Nordics, and the Grays. My speculation is that the gold is going to be used for helping the US and other national governments shift from a FIAT-based money system, which is predicated on Deep State-controlled central banks loaning governments money, to a gold-based currency system that will lead to the replacement of the US dollar as the world's reserve currency. According to economists, the shift from a FIAT-based financial system to a gold-based system requires that at least 5% of a national currency is held as gold. Otherwise, the national currency might collapse if there was a run on that currency. In short, JP's experience involving the international movement of huge amounts of gold supports the idea of an Earth Alliance battling against the Deep State, and that part of the battle requires shifting from FIAT to gold-backed currencies, and this requires a significant amount of gold being moved into secure vaults of national governments.

Chapter 8

Ant People Move to New Realm after Ant King Transitions & Sleeping Giant Awakens

On March 21, 2024, JP conducted another covert mission to an underground colony of Ant People, where he had previously met its King and saw human refugees whose ancestors had been given sanctuary centuries earlier. JP says that his latest mission was to discover what had happened to the Ant People who had disappeared, according to military reports. When JP and his four military companions entered the Ant People colony, they were greeted by humans wearing robes similar [to those] of the Ant People. JP's team was told that the Ant People had left for another realm, and the humans had remained behind to take care of the former colony.

When JP eventually met with some of the Ant People, he was told that the Ant King had transitioned and that a new king, who had just awakened, was now in charge of the Ant People, who had moved to a new realm. JP believes that the new king was the sleeping giant he had seen in an earlier mission to a separate, underground Ant People colony near Florida's Ponce de Leon Springs State Park, which appeared to possess a Tree of Life. It was speculated that the awakened giant was the mythical Ningishzida (aka Thoth, Quetzalcoatl, etc.), who was the Ant People's new king and now presided over a new realm that was located in a higher dimension. The transcript of the interview follows, with grammatical corrections and my commentary.[60]

Interview Transcript with Commentary

Key: MS – Michael Salla; JP – Pseudonym for US Army Soldier

MS - I am with JP once again. For those that are new, JP is an acronym, a pseudonym we use for someone who is currently serving in the US Army. We use that acronym just to keep his identity a secret, and he has been telling us about missions that he's been conducting, and he has been given permission to talk about these missions and disclose them. So, in a sense, he's not really a whistleblower. I guess this is like an unofficial sanctioned disclosure. That's how I can put it. So welcome JP to Exopolitics Today.

JP - How are you doing, Doc? I'm happy to be here again to bring this information to you, Doc. I know there's a lot of people interested in these types of missions and what's happening behind the scenes. A lot of people are really happy that this information is coming out in a way that they can understand.

MS - Well, wonderful. I know you have a lot of fans out there. A lot of people are looking forward to your updates, so I'm glad we can do another one for them. So yeah, what happened on your last mission? I think you told me yesterday (March 21, 2024) that you've just been on a mission. So, when was the mission, and do you want to tell us what happened?

JP - It happened actually yesterday, all day yesterday, and then we got back at 8 o'clock pm. We'd left early in the morning, and we came back at 8 pm. It was an awesome type of mission. We all got a text message in the morning to meet at this particular location in Florida, and we went to that particular location. We met with somebody that was in charge of the mapping. So, we were with him, and there was another guy with us as well. But he had a mask on. [He had a] really good build. We're all in good shape. We have to be in good shape in order to do certain types of missions. So it was a total of four people, four guys. And we had a fifth person, she was a girl, a female, and she was with us. She spoke a couple of different languages, so she was a linguist. Her name is Sue, and she was with us.

We were going to this [rural] type of location, right? So, we went to this [rural] location, it was a total of four guys and a female. We were going into the woods. So, we were following this path, we're heading north from our location, heading towards Alabama, towards that area. And it splits up to Georgia, so [somewhere] around that area. Today, we drove over there; we were all in a 4Runner Toyota type of vehicle. And we got to this location, but we met up in a place before. Then we got into this vehicle and started heading to that area. It ended up being the same area where we went [on an earlier mission], where I saw the bush. And I said, "Oh, man, I remember this bush."

The previous mission JP is referring to is when he visited the underground Ant Colony and met with its king in early January 2023. He described being taken to a remote rural location on a Mercedes white van, and they were met at the location by a helicopter that provided security. Once they were dropped off, he described another higher-ranking soldier, the "map dude," taking over the responsibility for land navigation (aka landnav). Landnav is an important skill set taught to Army personnel due to the possibility of modern GPS systems going down during a military operation.⁶¹ JP described how the most experienced person at landnav will become the "map dude" during a mission.

> JP - So the guy with the map, it was a different dude. It wasn't the same guy that came with us last time. He had a flute with him made of wood, a wooden flute, and he played it with a high-pitched sound. And the hatch shook the ground where we're at, we all shook, and it opened up. And I'm saying to myself, "All right, this is the same location, this is the same spot. Okay, so let's go in." A total of five of us. We're going down the stairs. This time, we had more lighting with us, and on the walls, we could see writings that we didn't see last time. So, we're like, "Wow, look at this." We were heading halfway down, and then we heard the hatch come down again, and the pressurized thing came up to our ears, and we started heading down.
>
> MS - But for people who maybe haven't heard this earlier story, I think a year ago you described going to that location, and I think you were taken there by helicopter, as I recall.

As already mentioned, JP was actually taken to the location by a white Mercedes van, not by helicopter, which was instead waiting at the drop-off location.

> JP - Yeah.
>
> MS - You were dropped off [in the previous mission], and someone opened that big circular or that big entranceway, and that's when you went in, and you went to the underground ant kingdom, where the king was.

ANT PEOPLE MOVE TO NEW REALM

JP - Yeah.

MS - Okay, so people who have kept up with your updates know where you are at, but this is the same location where a year or so ago you went down and saw the Ant King.

JP - So yeah, we started going down this location, and it's like a couple of [flights of] stairs down—10 minutes straight out again. That was the same way that we went, and we ended up in this open cavern place that I saw was familiar, and I saw two humans standing guard this time. I didn't see the Ant People. And I was like, "Okay, they were wearing the same robe that people wear." So, they nod their head, and we nod. The girl went straight to them and spoke a particular language that sounded like a Greek and Arabic mixture. They spoke back with her, and they just opened the door. So, this female was wearing a mask, but, you know, you could tell the difference between female and male. I knew she was a female, but her eyes were oriental, so she was, I think, Asian. Cool girl, she's about 5' 8" [172 cm], and she was with us, she's a linguist....

So, she talks to these two guards, and they open the gate. It's like a wall that opens up, and then it goes to the side, and it shakes everything. I think it's a type of hydraulics technology. I did see the bottom part of the door. It had like big huge marble round balls that's what it looked like…. These doors, when they slide, they roll over these tracks that have these big marble balls that [make] the door slide easily, but I don't know what pulls it to the other side but it does it automatically. So, it was quite interesting to see the construction of that door moving. I got an idea of how they move because I looked down, and I saw these huge marble balls and the big ass door. I think it weighed a couple of thousand pounds of granite or rock. It was just rolling on top of that, and I guess by leaning on a certain marble, metallic-looking marble, it could twist easily by using weight. So, it went in, and then it went to the side easily. You could just move it [with] one hand. It's just awesome construction.

So, we entered, and it looked quite different [to my last mission]. We saw again a lot of dogs, a lot of cats, a lot of domesticated types of hogs, and we saw humans. But I did not see the Ant People. So, I asked one of the guys. I said, "Hey, what happened to the Ant People." He says, "I don't know." So, I ran up to the girl and said, "Hey, the people that used to live here, do you know where they went?" And she's like, "Yeah, but that's what we're trying to find out. One of the reasons we're here is because we got a message saying that the Ant People left."

I'm saying like, "Why did they leave? We need to figure this out." So, we started walking in, and all I could see was humans. I did not see the Ant People. So that kind of worried me that I didn't see, you know, our friends. So, in the distance, we kept walking, and all we saw were humans dressed in robes, similar to how the Ant People were dressed. And I told this girl, "Hey, we were here a year ago," and I was speaking to her in a soft voice. I was like, "Hey, we were here a year ago. There were other beings here who call themselves the Ant People. Are they here, or are they not? What's going on? Like, what's the situation here? And there was a king that was here." And then she looked at me kind of weirdly. She's like, "a king?" I'm like, "Yeah, there was a king here. She's like, "Wait, wait, wait, this place hasn't been opened for like a hundred years. Nobody has been here." I told the girl, "What are you talking about? I was here last year, and I remember what I saw. I know what I saw." She's like, "No, this has never been open to the public. How do you know this?

I said, "I've been telling you everything since up there. How it opens, how we go down the stairs, how there's going to be two guards there." She looked at me like super weird, and the two guys behind me were puzzled. And I noticed that nobody, none of the same guys that were there on the first mission, was with me now. They are all different people, different map guy. We had three other guys, and then we had the female. So, I was thinking to myself, like, "What's going on here?" So, I just started meditating. And when I started meditating, I started seeing in the distance, like shadowy figures and the girl ... she's like, "What are you doing?" I'm like, "I'm meditating."

"How do you meditate when you walk?" I was like, "I just think about white light, and I semi-close my eyes, and I think about happy things that have happened in my life." Then she looks at the distance, and we start seeing these figures coming out of the caves. I started seeing Ant people dressed in robes, and they started coming out. I started seeing a lot of them [Ant People], and then all the humans that were there were kind of surprised, like they had never seen the [Ant] people. This was weird.

They used to live together. Now they're showing up again because we're here. What's going on? I'm not sure what's going on. One of the one of the Ant People approached us and was face to face with her, and she tried to communicate with them, but she couldn't communicate. And she looked at me. She's like, "Can you communicate with him?" I'm like, "Yeah, I can probably communicate with him." So, I stood in front of him, and I looked straight at his eyes, and he looked straight at my eyes, and I could hear a voice saying, "We went farther in," and I asked him ... "What do you mean you went farther in?" And they repeat the same thing. They said, "We went farther in into the Earth." And I said," Is there a reason why you went farther into the Earth?" And he said, "We can't disclose that information. But we left this part for the humans, and we went just farther in."

I said, "Can you take me to the king?" And he's like, "The king is no longer with us." So ... I started tearing up because while I knew the king was old, I never knew that he would probably pass away in our lifetime. So, I asked him again, "Is the king alive?" He said, "Yes, the king is alive, but he's no longer with us." I guess their belief is that when somebody dies or when something dies, they're still in the realm of the living, and they're still helping the living. So, I asked him, "How do you know he's still with us?" And he said, "We still feel him". He did this gesture, "We still feel him." I'm like, "OK, we still feel him."

So, I started thinking about it, about what was happening, about her telling me that this place hasn't been open for 100 years. But I don't think they knew about the last mission because it was a type of classified mission ... They use different people for different missions. And I knew I went down there, and I knew what I saw. So I said, "I need to know the location where the king is." So, he [the Ant person] went like this, and I went with him. I said, "Can they go with us?" He's like, "No, just you." So, the other guys, they stayed behind, and I went with the Ant guy. The Ant being I went with him. And I was right behind him. And I was just studying the way he was walking. The way they walk is like a limp, like a slight limp in each leg. They walk like ... they're dancing.

So, I kept following him. And I was smelling, like, honey... He was carrying a belt with pouches, and he had two pouches with him. And I saw these pouches, and I'm like, "Wow, what are these pouches?" So, I started talking to him. I said, "Hey, what are these two pouches that you have here?" He's like, "Oh, yeah, this is the grain." So, he picked it up, and he showed me, "This grain is what you guys call Chia." I'm like, "Wow, awesome. And then what is in that pouch?"

MS ... Like Chia seed, okay, so it's a drink.

JP - I did a video on my channel of a longevity drink

MS - Okay so this is the same drink you talked about in your video about the longevity drink.

JP - So I really wanted to put that video out before the interview. When this interview goes out, they can go straight to the channel and see how the drink is made.

JP created a video showing how to make the longevity drink based on what the Ant People told him which he released on his YouTube channel.[62] As he soon explains, it is a simple combination of chia seeds and pomegranate juice that are mixed together in a glass jar and refrigerated for 24 hours before being drunk over the next two or three days.

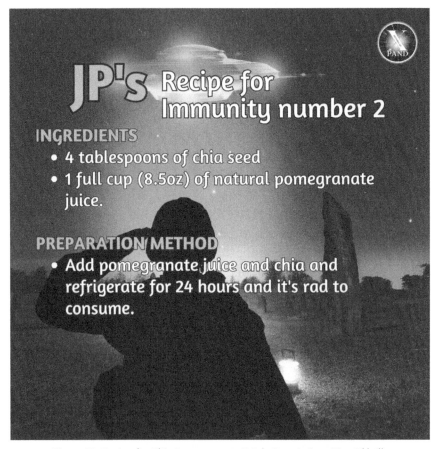

Figure 22. Recipe for Chia Pomegranate Drink. Permission: Dine D'ávilla

JP - But there's another type of liquid here, a drink that they showed me that I did not show on my channel, and that is going to be exclusively shown here. In this interview, people are going to learn about this other drink that they showed me. So, I kind of knew of the drink they were sharing with me, but it's interesting how they did it. So, he showed me the other pouch, but the other pouch was squishy, and I was like, "What is that?" "This is pomegranate juice." But they called it another name, "paragoness". They called it 'paragoness'. Their language is similar to Greek and Arabic, but I kind of understood what he was trying to say telepathically.

So, when he said, "paragons," I kind of knew it was a pomegranate. So, a similar word. He made me put my hand out, and he poured a little bit into my hand, and I drank it. It was pomegranate juice pure pomegranate juice. So, they showed me that they had these clay-looking jars, and they poured pomegranate in. Then they put chia seeds in these jars right that they put in a cool location. They wait 24 hours before drinking the pomegranate juice [mixed with Chia], and sometimes it ferments and turns into a wine, kabocha-style drink. So, it's really good for you, and he told me that it's good for longevity. So, I made a video about how to make that type of drink.[63]

JP - So, I kept following him, and I kept asking questions about the culture and the type of medicine that they drank. I also asked him his age. He said he was 86 Earth years. They say Earth years because they can age differently depending on what planet they're in. I did not know that. He said human beings can do the same thing. We can age differently depending on what planet or what moon we're on. The aging process that we have on Earth is exclusively for Earth. When you go to Mars or other planets, you age differently. If you're in space, you age differently. Your life can be longer, or your life can be shorter depending on the gravity, depending on what you eat on those planets, depending on the oxygen level, and depending on the microbial life forms that live [inside you].

Different planets also have different types of microbes that live there as well that also live on different extraterrestrials. ... This is like a protocol when you meet other ETs or if you go to a different timeline or a different time, of how to cleanse yourself before going to these different places. So that's awesome. I will probably do a video of that next time and explain how microbial life forms can affect ET contacts and teleportation and a lot of things that go into it.

A lot of people don't take into consideration that we do carry life forms in our body and on our body and that these life forms also travel with us when we do teleportation or travel to other planets and other places. So, whatever we eat can affect these life forms that help us. These microbes can help us live longer or shorter, depending on how you treat them.

MS – You said that the Ant person that's guiding you had two pouches, one with chia seeds and one with pomegranate juice, so they would just mix them and let them sit for 24 hours?

JP - Up to 24 hours, but some of them, they just eat it. They eat a chia seed as it is ... and they drink it with pomegranate. And what happens is that when it hits their belly, it grows, right? With the acid of the stomach combining with the pomegranate, it grows, and they don't eat as much as they're supposed to. So, it helps them with fasting. They don't get as hungry when they eat these chia seeds with pomegranate mixed in in their belly ... Now, when you drink it, and you make it after 24 hours, it helps you prolong your life. So, something about fasting and eating pomegranate and chia seeds, dry chia seeds, and then a drink of pomegranate can make you fast longer than the other. They also told me that they showed this to the Indians back in the days when they showed them about the pouches and all that, and how they can last all day with energy just using this type of method of eating.

So, it's quite interesting that they were showing me that. So, yeah, he showed me the room where they keep the clay jugs and the chia seed ... It had writing on it about how old it is. So, this particular Ant guy, he was around 80 years old. And he was healthy and strong. Their body is more defined than ours.

MS - I don't know if you said the last time you met the Ant people, but do you know how long they normally live?

JP - I know the Ant King was a couple of hundred years old, maybe like 400 years old, around there. He's been alive since the 1700s. So, it kind of surprised me when they told me that he had passed away, but he's still with us. I was wondering who took over. So that was my main question. Who took over as king, or who is in charge now? So, we started getting to that conversation and he showed me where the Ant King was laid. It was made of beautiful pearl rocks. They put pearl rocks around. It looked like a grave, but it was beautiful pearl rocks going around in the oval, and his crown was engraved with the rock. And I'm like, "Man, how did they do that? How did they engrave the crown into the marble like that?"

So, they have this type of mason work. How do you say it? Masonry. The type of masonry they work on is interesting, as is the construction. They know of the mixture of certain clay, and it's stronger than what we have right now on top of Earth. By the way, they know the knowledge that they have of mixing. Huh?

MS - Metallurgy

JP - Yeah, they got a lot of knowledge of how to mix things together and certain types of potions for longevity and diseases, and all that. So, I felt sad when I saw it. It was really, pretty sad.

MS - I remember the last time you said you saw ... some ships flying. The Ant People had technology, advanced technology, and flying technology, so did you see anything like that?

JP - We saw some ships here and there. They were not flying. I saw an orb that was going left to right. But like I said, again, there were a lot of humans. There were a lot of humans when I went there. I saw one Ant person. And he was showing me where that king was laid, laid to rest. And I asked him, "Hey, who's in charge now?" He said, "We have another King in charge, but he's not from this realm." I looked around and looked at the grave. And then I saw the hand of the Ant guy, and he pointed to a wall. So, I looked at the wall, and there was a beautiful artwork of the Tree of Life. I said, "Are you trying to tell me that the King who is now reigning is the King who was sleeping?" And he said, "I can't get into that information. You need to figure it out yourself." So, I went to the wall.

MS - But before you go on JP, when you say the king that was sleeping, you're talking about the giant, the sleeping giant. The very first time you went into one of the subterranean and kingdom chambers, you saw a Tree of Life, you saw a lot of Ant people, and you saw a big stasis chamber or sarcophagus with a redhead giant in it. And you were trying to find out when he would awaken and kind of that information. So right now, you're alluding, but the Ant person didn't confirm that whoever is in charge now is that sleeping giant, but who's awake? We don't know that.

On JP's first visit to an underground Ant colony in September 2022, he saw a "Tree of Life" that distilled rejuvenating waters absorbed from an adjacent stream and a nearby sarcophagus with a giant that appeared to be sleeping or in stasis inside of it.[64] The sleeping giant was revered and protected by the Ant People. Based on subsequent discussions with Elena Danaan and JP, we believe the giant to be the Sumerian god, Ningishzida, who was known as the god of the good tree, as I discussed in length *US Army Insider Missions 2*.[65] Ningishzida appears to have been known in other cultures as Thoth (Egypt), Quetzalcoatl (Aztecs), Kukulcan (Mayans), and Viracocha (Incas). Based on what JP learned, it can be surmised that "Ningishzida" has awoken and is now leading the Ant People.

JP - Well, he told me that he's from a different realm, but I kind of knew what he was trying to tell me when he pointed to the Tree of Life painting that he has on the wall. So, I went to the wall, and I put my hands on the wall, and I felt a vibration. I asked the Ant person, I said, "Hey, what is this vibration?" And he's like, it's the other side of the realm that we're not supposed to go and disturb. So, we felt like a humming vibration coming out from the other side of that wall. I said, is there a way that we can get there? He's like, "We can't go there. It's a different realm." And I went back, and I got the guys, and I got the girl, and the girl brought back like six of the humans that were living there. I started asking her to translate.

I said, "Hey, ask them if there's a way to get behind this wall." And one of the Elders, I think he was like, also, you can see in his face, almost 70 or 80 years old. He came up, and he had a long beard. He had sandals. He looked like one of those Romans with robes, and he had a staff with him. And he's speaking to the girl, he's like, "Yeah, I know, I know a way that we can go, but only one person can go through." And I told her, I said, "I'm not going without you guys. You guys got to be with me." And she's like, "No, you don't understand. You have to have a certain vibration to go into that location. It's holy."

He's saying that it's holy. And I explained to her, "Hey, I've been to a location where I saw a sleeping giant and the two guards that were guarding the, you know, sleeping giant. I saw the Tree of Life. I saw that realm. I know where that place exists. It's real." And she's like, "Hey, if you want to die, you can die. But I'm not going." And the other guys are like, "We're not going either. So, if you want to go, you can go by yourself." Well, I asked the guy to take me to the location where the entrance was.

So, he took me to the location where the entrance was. And there was a type of forest field on a square entrance. And you could put your hands in, and you could feel the hair of your body going up. And then, when you take it out, you don't feel it anymore. So, you could tell that this entrance was protected. And I was "like, okay, if I go in, what's going to happen? If something happens to me, it is what it is." So, I started walking into the square, and the guy looked puzzled, and he started speaking to me in a [strange] language, and I could understand him. And he told me ... "nobody has ever been this far." I'm like, "Okay." I kept walking, and my whole body started vibrating. And I can feel happiness going into the middle of my chest, going to my stomach, going to my throat, and then in between my eyes, behind my head, and my two eyes. I can literally feel them vibrating like this. As if my body was demateral, dematerializing. Did I say that correctly?

MS - Dematerializing.

JP - There we go. That's the word. Appreciate it, Doc. I close my eyes. I open my eyes, and I can see vegetation as far as I can see. I saw beautiful vegetation, and I could smell a beautiful, beautiful fragrance. Like if it was a mango mixed with guava type of scent. So beautiful. The place I saw, it and everything else was bright. Like, man, I hope I wasn't high or anything like that, but it looked like everything was really, really bright. And I thought about ... that drink that Ant guy gave me, you know, I hope it wasn't laced up or anything like that. You know? Because what I was looking at, the color was so magnificent. Like everything was brighter. And I was like, "Wow, where am I?" And I looked back, and it was dark. Whatever I crossed, it was not a portal, but it was a type of door that was different, one that dematerializes you and brings you back to another realm or place. So, I looked, and I saw Ant people. I'm like, okay, this is where they're at. I didn't know if I was farther into the location where I was. But I shouted, I said, "Hey," and they all looked at me like shocked.

I didn't see any humans, just Ant People. One came up to me, bowed to me, and said, "Why are you here?" And I said, "I want to know who's in charge of this location." And they said, "He that is not from this realm is in charge. He that has awoken." So, somebody who has woken up is in charge of this realm. And I said, "Can I see him?" They're like, "No, too holy." And they looked away when they said that. And I said again, "Can I see him?" "No, too holy." And I'm like, "Can he be a God? Can he be a certain person that's in charge? Or what is the spirit?" I'm like, "Okay, can I go farther?" He's like, "No, stay there. Somebody else will come." So, he leaves, and somebody else came.

And he was dressed in a whole robe, similar to the robe that the king used to wear. He's like, "I'm in charge of this location but not of the realm." So, you're the new king. He's like, "No, no, no, I'm not the king. I am in charge." I said, "Can you show me around?" So, they took me to this room, and this room was hollowed out and it had different types of candles lit it up around. And it had, like, a crystal in the middle of the room.

I looked at the crystal, and it was ... like a pinkish, purplish color type of crystal. And I sat down, and on the table, I saw a type of drink. I said, "What is that?" They said, in order for me to go farther, I need to drink this drink. I'm like, "Okay, all right." I asked them, "What is it made of?" [They replied] "It's made of ginger, garlic, and honey." And it tastes amazing because it also has a hint of lemon on it. So ginger, lemon, honey, and garlic. So, you take ginger, mush up the ginger, they mush up the garlic, then they grind it up. It has to be fresh. They get the honey and pour it like a type of lemon. It didn't look like a lemon, but it looked like a lime that we have. It was a mixture of green and yellow. It was a type of lemon, and they squeeze it in and pour it around ... and they make you drink this beautiful drink.

And I drink it. And the moment I drank it, my whole sinuses cleared up. I felt energy that I hadn't felt in a long time. And I start coughing. And I asked, "Hey, why am I coughing?" Like you're coughing, and it helps you release everything that you have. So, I started sweating. I started feeling better. I sat down and said, "I need to go back. I need to go back. I need to go back." So, what I felt in me was that I needed to go back because I was feeling good, but I didn't think I was prepared to see what I was supposed to see. So when they started taking me to this corridor—it's like a long, long, long tube—like a tunnel that they built, I started looking at the walls and all that. I was like, "I need to go back." I started pointing back, and they were like, "You want to go back?" And I'm like, "Yes, I want to go back." So, it's not that I felt no peace. It's just that I don't think I was ready to see what I was supposed to see, Doc.

So, it's nothing against them. Nothing negative. It's just I needed to go back.

I felt that I needed to go back. What I feel, I follow what I feel. So, if I felt I wasn't ready to see what they wanted to show me, I was honest with them. And I told them, "I need to go back." So, I went back to the room. They made me drink again. And I went back to the same location. I came out of the square, and my whole body started vibrating again. And when I came out from the other side of the door where the guys and that one Ant person was and the female was, I started throwing up. And I was throwing up like a bluish substance. I don't know what it was. But then the Ant guy told me, "Yey, you're healed." I'm like, "What are you talking about?" He's like, "You're healed." So, everything was falling kind of into place. What were they trying to do with me? Probably to heal me in a certain way for me to keep on doing what I'm supposed to be doing. And I know that the Ant king told them about me and what to do with me next time I came here to that particular location.

So, we started heading back, and the Elder guy that had the beard with the staff, he said, "We want you guys to come back here. And we're going to show you other things that will surprise you guys. And soon we're going to share with the surface people." That's what the king said, "to share this information with the surface people." But we were like, "Okay, is it going to be every location in the United States and every location around the world that you guys are going to share this information?" And he's like, "Yes, if the king from the other realm blesses us to do so." So, I guess the other king from the other realm was

> developing something that was going to help us move forward in our time in our realm. So, it was an amazing experience, doc. We started heading back.

What JP just shared is highly significant for the coming years as humanity becomes reacquainted with the Ant People and other Inner Earth civilizations. This will facilitate the sharing of much-suppressed knowledge, such as the healing techniques using different plants and advanced technologies. A recently published book in German reveals that up to the 1500s, there was open communication and contact between surface humanity and the Inner Earth.[66] However, after major UFO sightings and battles over Europe in 1561 (Nuremberg) and 1566 (Basel), the Vatican led an international campaign to stop all contact with the Inner Earth. All tunnels and entrances to the Inner Earth were sealed, and people were forbidden to talk about Inner Earth civilizations. What JP was told suggests that after more than 460 years, open communications and contact with the Inner Earth are about to resume.

> JP - We came out from a different location. We got picked up by a truck, like a GMC truck. We got taken back to our location where the bush was. We got into the 4Runner, and we started heading back to where our cars were. And I came home. Coming back, I got ginger, I got garlic. And if you do this recipe, try to get the honey from your location, the honey that is in your location, no other location. Don't buy honey from the store. It has to be a natural honey from your location in order for this to work as well. Because that's what's in the air you're breathing. That's the pollen that is falling. Everything counts together.

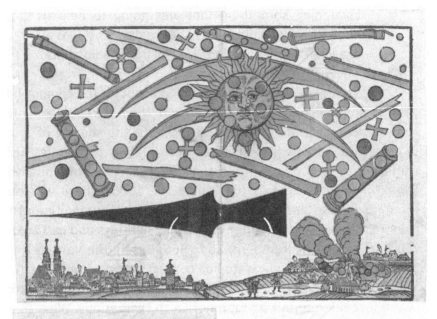

Figure 23 Celestial phenomenon over Nuremberg, Germany, on April 14, 1561. Illustrated by Hanns Glaser

JP - So, you grab ginger, garlic, honey, and lime. And you mix it in together. You put it into a similar jar where I showed everybody how to make the chia and pomegranate drink.[67] And this particular drink is going to help you with colds. It's going to relieve the arthritis pain. It's going to help your brain to think better. It's going to give you energy if you drink this in the morning. And it's going to cleanse your body with any type of phlegm that you have built up. MS - So this happened yesterday ... March 21st. Today is Friday, so this happened on Thursday. So, it's only been one day, but you had that drink, and they said that you were healed. So, do you feel better? Do you feel that this has helped you, that drink that they made for you?

JP - Yeah, I feel good. My first day trying it, you know, I did feel good drinking this drink, the liquid substance that comes from ginger, garlic, honey, and lime.

MS - Now, what do you think the Ant person meant when he said that the Ant people have gone to another realm or that they've kind of moved from that location that's been handed over just to the humans now? And they said that they had moved to another realm where this king of the other realm presides.

JP - It's a higher vibration place. So, it's a higher vibration where they go that I don't think a lot of humans can go yet.

MS - Okay, so we're talking about a fourth density, fifth density, and vibrational realm that they've gone to and that this king who's awoken has somehow opened the doorway to that realm for them to go into.

JP - Yes. Exactly.

MS - And you haven't confirmed this, but you suspect that it's the same sleeping king you saw in that sarcophagus a year and a half ago?

Figure 24. Recipe for Immunity given to JP by Ant People. Permission: Dine D'ávilla

JP - Yes, I suspect that was him. That's why I think I wasn't ready to see him. I don't think I was worthy yet to approach him, nor was I sure what to say or how to talk. So, I wasn't feeling comfortable with

ANT PEOPLE MOVE TO NEW REALM

myself. Maybe for them, I was ready, but for me, I just felt, "I'm not ready."

MS - OK, so the main mission was to find out what happened to the Ant People, and I guess you succeeded in that mission. You found out that they've simply moved to a higher realm or another realm, a density realm where they simply, as you said, appeared through the cave walls. So it sounds like they are in another vibrational realm where they can just shift in and out of our reality when they want to.

JP – Yeah, I agree. They looked like shadowy figures instead, but when you approach, and you go through the door and you go through the square vibrational part, you can physically see them. But the place is like maybe holy, I guess. It's beautiful. It's bright. The colors are brighter. I don't know if the air is made of marijuana. I don't know what the heck is going on, but I can't do drugs. I can't do any of that. But you feel happy. You feel good. You feel awesome and in a beautiful place—feel peaceful.

MS - So those two drinks that they gave you and explained to you what they were, that sounds like they are part of the process of being able to move from a third-density realm to the realm that they're in right now.

JP - What do you mean by process? Are they part of the process of moving to that realm?

MS - The drinks that they gave you, taking those drinks, do they help a person move into the other realm, or is their consciousness more capable of dealing with these higher density realms?

JP - Yes, it helps with the cleansing of the human body, the cleansing, and helps the microbial life that lives in your body to have, I guess, a safe sensation that you are feeling, something like that. It's a weird science, but it's interesting if you get into it.

MS - Now you said that the Ant people were going to start showing themselves to people all over the Earth. So, does that mean that they're going to come out of these higher-density realms and just appear on the surface and interact with people? Because they look different. They don't look like humans. So, how would that work?

JP - I think they would find a way to communicate with humans. And the humans that live there. They could actually even walk among us, coming out and living and walking among us the same way that the Nordics do. I found that quite interesting that these humans, they really look human. They're all human. And the Ant people, yes, they do look different, but I think, yes, they can show themselves to certain people who are ready for them to be shown. I think they can appear from different places, depending on where you're at. If the area has a higher magnetic field or it's a place that has been holy before, maybe they can show up there. They're going to start showing themselves to certain people. And people are going to talk about them like they used to talk about them.

As I previously mentioned, there is evidence that communication and contact with Inner Earth civilizations was openly conducted until the 1560s, when major UFO sightings and battles led to the Vatican intervening. In his book, *Inside the Earth:*

The Second Tunnel, the author, Radu Cinamar (a pseudonym), discusses historical communications and contact with Inner Earth civilizations.[68] He believes that tunnels and entrances were sealed and that today, they are closely guarded and monitored by both Inner Earth civilizations and secret societies such as the Freemasons.

> MS - Well, of course, it's going to be very selective. It's going to be high-vibrational people who can.
>
> JP - I assume it's similar to the Sasquatch people, you know, a similar type of way.
>
> MS - Okay, so it's just like the Sasquatch, you have to be in a certain vibration, a certain way of thinking, and like in a place of love and empathy, and they'll reveal themselves that the Ant people will be the same. It's like you've got to raise your vibration so that it can match with them, and then they'll reveal themselves.
>
> JP - Agreed.
>
> MS - And so you said that that's going to happen all over the world, so there's no one place you go to. It's just that you've got to feel guided. It's probably going to be more in nature, like in the woods and forests.
>
> JP - Yes. That's why I've been putting videos out encouraging people to go into the forest, go into the woods, and really connect with the wildlife and connect with nature. Because that's the only way that you could feel these types of things connecting and then having that extremely big love, and I show a lot of love. I try to show as much love in the

channel, saying I love you and I love everybody who follows me and cares about this information coming out and touching their hearts by sharing information on how to get closer to ETs and how to get closer to them by what you eat.

JP - I'm noticing now that I'm getting traction with people, and I'm not being threatened. Look, I'm not lying. This is right now. I'm talking about this. Hold on, do you hear me?

MS - I hear you, yeah.

JP - So right now, a Black Hawk [helicopter] just flew over my head.

MS - OK, and you're at home, so that's unusual.

JP - Yeah. So let me just stop there, yeah.

MS - Okay, any final words you want to say to our Spanish and Portuguese friends?
JP - ... My grammar is off the chain sometimes, but I just hope you guys get this information and have an experience when you go into the forest, or you go to the woods and you do a meditation, I wish you a beautiful experience that you guys may have contact in a spectacular way. More people are gonna have these experiences. I'm getting a lot of people Doc and in the social media, a lot of people that I sometimes that by asking a particular question that I know how it is and they have been through similar experiences of visiting certain locations and having contact with different types of things and communicating with them.

So, there's gonna be people having these experiences, and they have to be open about it. And if they can text you or send you an email, you know, on your website, because there's gonna be a lot of people going through different types of experiences. And I can't wait till you or other people go through these special experiences. I know you probably had experiences here and there, but I think it's gonna be more intense with people, the experiences that we're gonna go through.

MS - I think you're right. Well, I want to thank you, JP, for again giving people this update. It's always great to hear about these missions, and I know it kind of takes something out of you every time you do this. And there are always risks, so we do appreciate it.

JP - Yeah. I do feel a type of sadness that the Ant King has passed, and it's really sad, but they're in a better location, I tell you that. The Ant people are in a better location, and they're going to be showing themselves soon to people.

MS - Wonderful. Well, thank you, JP.

JP's mission to the Ant People colony suggests that we will hear more about what is happening in the Inner Earth as they have a new leader who will decide when the right time has arrived for open contact to occur. If the Ant People's new leader is Ningishzida (aka Quetzalcoatl et al.), it's likely that open contact will be coordinated with the return of the positive Enki/Ea faction of the Anunnaki and the appearance of Nordic extraterrestrials (aka the Galactic Federation of Worlds). This means that open contact will involve the sharing of both alchemical healing sciences that

Ningishzida specialized in, but also advanced space technologies that other Anunnaki possessed and which the Nordics are helping human pilots to master, as we'll see in the next JP mission report.

Chapter 9

Nordics Training International Military Pilots to fly Flying Saucers

On April 15, 2024, JP traveled to an underground spaceport somewhere in the US state of Alabama, which appeared very similar to or the same as the one he visited in May 2023.[69] He again saw fleets of saucer-shaped spacecraft being piloted by human-looking 'Nordic' extraterrestrials. These flying saucers were identical to a spacecraft he witnessed and photographed in 2018 while living in Orlando, Florida.

JP says that he and another witness were accompanied by 30 military pilots from many nations. When they arrived at the underground location, all the pilots left and walked over to an assigned spacecraft and the Nordics each craft was associated with to continue their training. The human pilots were being trained to use the advanced mind-technology interfaces that are part of the Nordic spacecraft. Finally, JP was told that the Nordic spacecraft flown by military pilots from different nations would soon start showing themselves in a worldwide disclosure initiative. The transcript of the interview follows, with grammatical corrections and my commentary.[70]

Interview Transcript with Commentary

Key: MS – Michael Salla; JP – Pseudonym for US Army Soldier

> MS - I am with JP. It is April 16, and I want to welcome you back to Exoolitics Today, JP.

JP - How are you doing, Doc? I'm proud to be here to bring this beautiful information to the public.

MS - That's great to hear that you've had another mission and you're being given the green light to share that with us. So yeah, tell us what happened to you.

JP - We went out from where I work. This happened around 5.30, 6 am in the morning. We all met up, a total of four of us. And we were going into like this bus, right? And this white bus picked us up—a 44-passenger bus. It picked us up, and we went into the bus. We were dressed in civilian attire, and we were confused about where we were going. We were heading north back up there to Alabama and Mississippi. I noticed that we were picking up people on the way. And I noticed that there were pilots, they were officers, different types of pilots, and even pilots from different countries. They told me not to talk about which countries were part of this type of exercise. This type of mission was more like a training type exercise.

So, the bus started getting filled, full of people, you know, everybody was quiet. Then people started talking in the back, and the wave of everybody talking started. Talking about, like, "Where you're from, why you're here?" You know, everybody had an idea, but I personally didn't know where we were going. There were a lot of people, like, a total of 32 in a 44-passenger bus, and we drove off to north Alabama in that area by Mississippi to a cavern.

Northern Alabama is famous for its large limestone caverns, and there is a popular cavern trail visited annually by hundreds of

Nordics Training International Military Pilots to fly Flying Saucers

thousands of tourists that spans the length from DeSoto Caverns in Alabama's south, to Russel Cave National Monument near the city of Bridgeport in the north.[71] This area of northern Alabama is also where the famous Huntsville NASA Space Center used for the Apollo Program is located next to the US Army's Redstone Arsenal, which is where Operation Paperclip German scientists were first taken after World War 2. German Paperclip scientists played a leading role in the Apollo program and the US nuclear missile program. In my 2018 book, *Antarctica's Hidden History*, I discussed how the Apollo Space Program was an elaborate cover for massive funds to be siphoned off to Antarctica and South America as part of a secret agreement to expand a German-run space program in exchange for the technological know-how for creating a US secret space program.[72] Large underground caverns in the Huntsville, Alabama region would have made it easy for this agreement to be implemented without the general public learning what was happening.[73] This makes it very plausible that a large underground spaceport currently exists in the Northern Alabama area that JP visited.

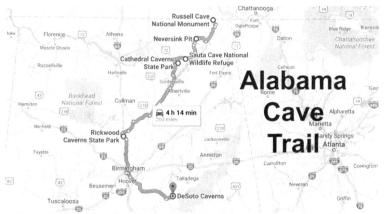

Figure 25. Alabama Cave Trail shows extensive caves in Northern Alabama

JP - We enter into this tunnel. It was really, really tight. I don't know how the bus driver took it in there, but if you could put your hands out of the bus,

235

you could feel the rock on the side. It was like three feet on one side [to the rock wall] and three feet on the other side. That's how tight it was. By making a slight turn, the bus actually feels like it's scraping [the wall] a little bit. And the guy that was driving was like, "Oops, sorry, sorry." Everybody laughed when the guy did that. It was really interesting. So we go into this cavern, and he parks in this big, big place. I did notice the bus going down into [some] type of ramp, and everybody there was excited. Like, everybody knew, and me personally, I did not know, but it looked like the other people knew what was happening. So, when we get to this cavern, and we get out in the distance, there's like a tube-like cave that you walk through, and they had lights [on], and we had to pass through these doors made of metal. These doors had insignia on them, and the insignia looked like a flying saucer. On top of that, they had another language that I did not recognize. I did not recognize the language or the writing on top of this insignia that looks like a UFO.

The door opens, and then we enter one by one. I did not know anybody who was there. I think I remember the driver. I had seen him a couple of times here and there, but I didn't see any other people around that I knew. I noticed people talking in different languages. I did not recognize different languages, something like Arabic or maybe Hebrew. I'm not sure, but there were other languages there as well. There were other pilots there from different places. I guess they decided that this was for everybody to meet up and go to this place. So, I asked one of the guys behind me, and I said, "Hey, I'm not a pilot. You know, I have flown [in] planes here and there, but I'm not a pilot." "You know," he said, "I'm not a pilot either." So, I kind of got confused because the people that we were picking up were mostly pilots. It was a mixture of different types of people that were there.

And we entered this place where we're going one by one. And then in that room was that door, that metal door that had the insignia of the flying saucer. I entered [through] it, and my whole body started vibrating. Then it stopped, and you could hear an intercom type of voice. "Okay. Keep walking, keep going, keep going." So, it was guiding us to keep going. So we entered into this big, big cavern with flying saucer ships already floating.

And they were lined up similar to the port that I went to [in an earlier mission]. I think this was the same area, but it was in a different part of this ... [massive] base, but these ships were high up. They were at least 20 to 30 feet high, and they were floating and humming. Zoom, zoom. And they were all humming ... all at the same time. Sometimes, one of them hummed differently, and then it entered into the same frequency, the same hum. Under these ships, there were Nordic guys aligned to each ship.

And then the pilots, they were running to each Nordic that they knew. So there were at least 30 to 40 ships that I was looking at, that were floating. And I asked the guy next to me, "Hey, what's happening here?" He's like, "Oh no, it's the type of training that they do. They train the pilots to fly their ships." And I'm like, "Wait, what do you mean they train the pilots to fly their ships? "So yeah, the Nordics, they take the pilots, and they're training them how to fly these ships." And I said, "To train how to fly these saucer kind of ships?" He said, "Yeah, yeah, they have to be connected with their Nordic, and the Nordics, they train them how to use it." So, I was like, "Wow, it is exciting to see this."

So once the pilot approaches the Nordic being ET, the Nordic raises his hand, and the ship starts coming down really slowly. And when it's like on top of their heads, they just disappear, they go into the ship. They disappeared so quickly into the ship that I couldn't see any entrance or anything. But the pilot was holding on to the Nordic, was holding on to his wrist. When the ship came down to their head, they just went into the ship. So, one by one, you can see the ships picking up the pilots and leaving the cavern system.

And then, while they were leaving, I could see the ship turning into different colors and mimicking the same color of the cavern. They kind of disappear. And I'm like, "Oh, this shit is using like a type of cloaking style [technology] of mimicking the cavern. That's awesome." So, it actually looked like a mist. It turned into like a moving mist. It was quite interesting looking at that and looking at the ship doing that because it made me think, "Man, how many ships are out there that we can't see?" And they're moving around and we just can't see because they're mimicking the clouds or they're mimicking the blue sky. And they just look like mist, like a heat wave ... When it moves, it looks like a heat wave mist. It's incredible. It's amazing.

MS - So I just wanted to confirm, okay, so this happened on Monday, April 15 [2024] this mission, and you're pretty sure that it was the same cavern complex in that Alabama region where, I think, it was about a year ago, you described going to this giant spaceport and all these Nordic ships, and they would go up through the mountain, that it was underneath the mountain, and that they would go up through the top of a mountain where there was a light beam or a holographic light beam that would disguise ships entering and leaving. Okay, so that's the same base, or maybe you came into it from another entrance.

JP - Yeah. We came in from another entrance, but yes, heading over there was the same way. So, yeah, I'm sure it's the same base but in a different location. Okay, a huge, huge base.

JP's earlier mission into the same base happened a year earlier, in April 2023, and was discussed in *US Army Insider Missions 2*.[74] He said he saw hundreds of saucer-shaped ships in the spaceport and that many extraterrestrial civilizations were involved. JP also described many orbs being released from the underground base to monitor surface humanity, and that the saucer ships would be increasingly flown and unveil themselves so surface humanity could acclimatize themselves to such advanced space technologies appearing in their skies.

MS - Okay, and I think last time [during] that last mission to that underground base, I think you said that these were Nordic craft and that they were going to start revealing themselves …

JP - Yes.

MS - Okay, so it seems like this mission that you just did on April 15 [2024], it's like, this is where human pilots are going to be part of those ships going out there revealing themselves. Is that pretty much what the agenda is here?

JP - I believe so, Doc. I believe that the time is going to come when these ships are going to reveal themselves to certain kinds of people. Or maybe the news or maybe to try to stop something that's happening.

MS - You mentioned that these pilots were kind of from all over the world …

JP - Yes, but there are more United States pilots, of course. Yeah, there were more United States pilots training with these types of Nordic beings. They were wearing uniforms, but everybody [else] was in civilian attire. But the Nordics, they're wearing the Air Force type of uniform that I saw … [on] that particular Nordic over there in Orlando who made contact with me and asked me to go into the ship. And I said, "No, I don't feel comfortable going into the ship." So, I declined, and he left. That's when I took a picture of that particular ship that was flying. And that's the same type of ship that I saw in these caverns, the one that is reflecting the trees at the bottom.

The mission JP is referring to here happened in 2018, prior to JP joining the US Army while he was living in Orlando, Florida.[75] JP was telepathically summoned to leave his home at night and travel to a nearby wooded, secluded area where he met a Nordic wearing a USAF uniform with a distinctive patch. The Nordic's hair was crew cut style and the insignia on his uniform designated a special unit within the USAF. JP declined to enter the ship because he felt uneasy due to him being monitored by unknown individuals, likely linked to the CIA. However, he was able to take a photo of the departing spacecraft. [see earlier figure]

> MS - Okay, well, that's important because you actually have photos. People always say, you know, where's the photos proving any of this? You actually have a photo from that 2018 incident where a Nordic invited you to go into the ship. Okay, so these are the same ships, they're wearing the same uniforms, that this was a faction of the Nordics that works with the Air Force
>
> JP - Ahem.
>
> MS - The human pilots were of different nationalities, so presumably, some of those ships would just fly over the US, while other ships might be flying over the Middle East or over Russia.
>
> JP - Or maybe they'll fly straight up into the spatial area [space] and fly to whatever country, whatever place that they desire.
>
> MS - OK, all right. So this is like a coordinated international effort where all these pilots are being given exposure to this American spaceport where there are all these Nordic flying saucer craft, and they're given a chance to learn how to fly these craft and collaborate with the Nordics. So it sounds like a very big, important multinational mission.
>
> JP - Yeah, it's not like a mission. I think it's more like training. It looks like they've been doing this for a while. I just haven't seen it, but it looked like a training type [of mission] ... the way they're doing it, because the pilots, they already knew the Nordics,

they walked up to the Nordic, the Nordic raised up their hands, the ship comes down, and they disappear into the ship, and the ship flies away.

JP - So they've been doing this for a while. I just happened to go and see it with my own eyes. But I think it will be quite exciting when people start reporting these particular flying saucer-looking ships. And also, on the side, I saw a ball [bunch] of orbs coming out from these these ships. And I asked the guy, "Hey, do you know what these orbs are? What do they do?" So, what these orbs do is they come out from the bottom part of the ship, and they analyze the area before the ships go. These orbs, they go and analyze and detect any danger for the ship. And it could be any electronic or nuclear type of radiation or anything that bothers the ship for it to communicate. Because, you know, the ship is like a biological machine. It's alive. It is like a type of artificial intelligence that surpasses anybody's mentality or thinking technology [AI].

So, the ships themselves send these orbs out to investigate the area where they go prior [to leaving]. They send out the orbs first, reconnoitering, and then they go exactly to the spots where these orbs go. And I think these orbs have the ability to transport ETs to certain locations as well. So it's quite interesting. The tool, they have these orbs that seem also really, really smart, connected to the ship.

JP's reference to the orbs being sent on surveillance missions before the flying saucer craft leaves any earthly location is very significant. On May 31, 2023, Dr. Sean Kirkpatrick, the former Director of the All-domain Anomaly Resolution Office (AARO), revealed: "The vast majority of the reports ... are sightings of unusual orbs or round spheres."⁷⁶ This is consistent with JP's observations that large numbers of orbs are being sent to reconnoiter different regions of Earth to ascertain the safety conditions for flying saucer craft leaving from the underground spaceport in Alabama and elsewhere around the planet.

Another important point raised by JP is that the orbs can carry extraterrestrials. On September 5, 2015, JP says that he was picked up by a very bright orb and transported to just outside Earth's orbit.⁷⁷ He said he was then joined by two of the same Nordics that he first met in Brazil in 2008, and the orb took him and the Nordics to somewhere in the region of Jupiter before JP he was returned home two hours later.

> MS - OK, so the orbs are playing a kind of monitoring, surveillance role to make sure everything is OK around the ship. Now, you said that there were 32 people on the bus, and most were pilots. So, how many of you actually were just watching the pilots go to the Nordic spacecraft?
>
> JP - There is one guy behind me, and that's the guy who said, "No, I'm not a pilot." And that was the guy I was asking the questions and asking, "Hey, what they're doing?" He was telling me, "They're training, they're, they're flying the ships." That could probably explain why they crash so much. They're so advanced that I guess the pilots can't comprehend how they work sometimes. And I guess sometimes they do crash. So, I think the Nordics are really putting their ships on the line when they hand it over to these pilots.
>
> MS - Did that guy tell you what the training exercise was all about? Did he say anything about how the human pilots are able to fly the ships?

JP - No, he wasn't telling me, but I remember when I entered into one of these Nordic ships, you know, that ship that I entered where I live. It can read you. It can feel you. And I guess you project it out of where you're going, and it goes where you project it to go. So it kind of knows where you want to go.

JP is here referring to the Nordic flying saucer he entered on November 4, 2023, that landed near his Florida home. The Nordic, Jafis, offered JP a blue drink, which he drank and made him feel better, as discussed in Chapter 3.

MS - Okay, so kind of like interfaces with your mind. So it's a mind technology interface. You would have to have a pretty disciplined mind. It would take a lot of training to be able to fly one of those ships. You know, you couldn't just go there and say, okay, you know, [I can fly this].

JP - Oh yes. I think these pilots have already been out there in our solar system. I heard them on the bus talking about different parts of our solar system, talking about Phobos. They were talking about flying around certain parts of Venus. These pilots have been around and they pilot other types of ships as well. So, it's not just these ships that they pilot. They have a history of piloting other different things [craft].

MS - Okay, that's very interesting. Elena Danaan talked about this Singaporean super soldier, Stephen Chua, who went, I think it was sometime around 1980. He went to Area 51, and he was trying to help Air Force pilots fly some of these fly-by-thought variations of them. There was an F-15F series that had fly-by-thought technology, but they suffered brain damage because their brains weren't ready for it. They weren't developed or disciplined or trained, whereas Stephen Chua, he had plenty of gamma brainwave activity, so he could do it easily, but the American pilots or the other pilots couldn't do it without a drug, and that drug just messed up their minds.... Steven Chua said it was around; I think it was 1980 that that happened. That means that in the last 44 years, this kind of fly-by-thought technology has been mastered.

Elena Danaan discussed Stephen Chua in her 2023 book, *Area 51*.[78] In it, she discussed his extraordinary fighting abilities and high intelligence that led to him being recruited for a special covert unit with the Office of Singapore's then Prime Minister, Lee Kwan Yew. Chua had unusually high gamma brainwave activity that came to the attention of US officials, who persuaded the Singaporeans to loan him to the USAF, where he was taken to Area 51. At Area 51, he easily flew the modified F-15F due to natural gamma brainwave activity that could not be easily replicated using technology or pharmaceuticals. This led to the abandonment of the hybrid flying program at the time. Chua was also a supersoldier trained to fight against different types of extraterrestrials on behalf of select governments that would send him around the world.

> JP - Well, I believe they tried to mimic it to put it on our ships, our jets, and our different types of flying, like TR-3B-like ships. I guess that it didn't work out as well as when the humans fly the Nordic ships, that … [have far] surpassed the technology that we have. Yeah, reading the brainwaves, reading your thoughts, and I think you have to be in a state of mind, yeah, like you said, like a trained state of mind. I think that's why they fly with the Nordics, you know, and if they pass out, the Nordics take

over, and I guess they come back, and the training is done.

It was really, really interesting. I was looking around the port more, trying to see more details, and I saw different languages, and I was seeing pictures of different people that have been there, that they have on the walls, and it looked like more military, certain politicians. I'm sorry that I didn't get the green light to share who the people are that visited these places already. They're big names. I think everybody would know who if I said the name, but yeah, big people have visited these types of ports, so it was quite interesting seeing that wall.

MS - Can you describe the languages you saw? Are we talking about modern languages, or are we talking about ancient ones?

JP - Yeah, modern languages. I think I saw Japanese. I saw a type of Greek language, Arabic, Hebrew. I also saw ancient writings as well, Aztec style. What I was told is they also have Inner Earth people training, but I think they come from a different time and not the same time as the humans. So, I believe other types of people come there to train as well to fly their ships.

This is a fascinating revelation made by JP. Apparently, some of the Inner Earth civilizations also require training to fly the advanced craft possessed by the Nordics. While JP has seen some of the antigravity craft possessed by the Ant People and other Inner Earth civilizations, these are presumably used for traveling around Earth, to the Moon, or for interplanetary travel in our solar system. The Inner Erath craft, however, is not as advanced as the interstellar craft used by the Nordics.

MS - So, do you know anything more about these Nordics? I mean, are we talking about Nordics that are based on the Earth assisting the Air Force?

JP - They're based both on Earth and outside of Earth [on other planets and moons], and these are the Nordics that are working with certain militaries around the world and also are in charge of some arks that we talked about in the past.

MS – Okay, all right, so ... you don't know anything about them being connected in some way with the Galactic Federation of Worlds. You don't know of any connection with them. This sounds like it might be a completely different group altogether.

JP - Negative, negative. I haven't heard of that yet. I haven't asked. They're probably involved, but I haven't asked. I don't like to try to ... what do you call it? ...

MS - Identify them?

JP - Yeah, I don't want them to think that I'm trying to, how can I put this, like group them, you know. Yeah, categorize them, or you know, [ask] who they are, If they are working for the Federation. Yeah, I don't know that. I have to ask that, I guess, the next time I see them. I will ask, that'll be a good question to ask one of the Nordics to say, "Hey, are you connected with you know? With the Federation and all that? Yeah.

MS - Yeah, that'd be good. And what star system do they come from?

JP - I think there are more nations that are involved with the Artemis Accords that are involved in the flying of these ships. That's the nations [of the pilots] I saw.

MS - Oh, okay, that makes sense.... There are 36 [currently 43] nations that make up the Artemis Accords. So, if you sign up for the Artemis Accords, then some of your pilots get to be trained. And it's interesting. I think 36 nations have signed, and you said there were 30 pilots out of the 32 people on the plane. So, kind of like ...

At the time of the interview, there were 36 signatory nations to the Artemis Accords. By August 2024, this had increased to 43. If military leaders from major spacefaring nations such as Brazil, Germany, India, France, Japan, etc., are told that their pilots would get to train and fly on Nordic-supplied antigravity spacecraft out of secret US spaceports, this would be a powerful incentive for national political leaders to join the US-dominated Artemis Accords.[79] According to Elena Danaan, US leadership in space was endorsed by different galactic organizations at the Jupiter Accords held in October 2021.[80] In contrast, the China and Russia-led International Lunar Research Station Initiative had only 12 signatories by August 2024, most of which are countries with small or poorly funded space programs.[81]

JP - Doc, I think they're the best of the best, by the way. That's how they were acting, like jock-type of dudes. I think they're like the best of the best. I felt like I was in a [Top Gun] Maverick movie.

MS - Okay. All right. So, there were 30 guys, 30 pilots on the bus that got off and went to the ships, and just the two of you stayed back and watched it all.

So, what happened? You just kind of watched and looked around while they did their outer space mission, and then they came back.

JP - Well, they left. I did not see them come back. But then they took me to a different room. And that's when I saw the orb. You could clearly see the orbs coming out. They look like lightning, and they come out in groups of two. They just look at different spots, and then they go back to the ships. They were timing it every 30 minutes. These orbs, they come out, and it's kind of cool because they go through walls and shit. I'm like, wow, OK, they go through walls, and they go through like certain types of rock. And I could see a crystal-like cave system, you know, and I could see like crystals. These orbs went to these caves and these types of crystals. When they get close to the crystal, the crystal lights up, and then the orbs leave. It goes back to the ship, and you can still see the crystal lit up because of the interaction with the orb, and then it dims down. So, I guess it was grabbing energy from the orb, or the orb was grabbing energy from the crystal. That was quite interesting.

I think they're really, really hot. I had an orb get close, at least 10 feet from me. I could feel the energy of the orb, an electric type of energy that I felt in Brazil when I went close to a ship and touched it. I started feeling the metallic [taste] on my tongue. My hair is going up. And these orbs were just 10 feet away from me. I think the energy that carries it can kill somebody if you touch it directly. I guess when it interacts with you, it doesn't kill you. But if you approach it without it interacting with you, I think it could kill a person with the energy that it has. It has to approach you. You can't approach it.

MS - Yeah, I remember that was something Tucker Colson talked about where he described a Stanford medical researcher, probably Dr. Gary Nolan, saying that he had treated about 100 Air Force personnel who had suffered traumatic brain injury or death when they interacted with UFOs. So yeah, that definitely is what happens. Okay, so you saw the orbs. So, what else did you get to see in that cavern complex while you were there?

In March 2023, Tucker Carlson appeared on a podcast and described a conversation with a Stanford University Medical Professor (likely Dr. Gary Nolan) who was tasked by the Pentagon to investigate military servicemen dying from brain injuries during UFO incidents. Carlson said:

> We got a call from a guy who's a tenured Stanford University Medical School professor. He comes on, and he's like, "Eleven years ago, the US government reached out to me because I'm an expert on head injuries, on brain injuries—traumatic brain injuries—and they had all these court cases from families of the US servicemen, over 100, who'd been killed by UFOs.[82]

Carlson's revelation corroborates what JP was told about orbs (UFOs) being dangerous if approached without permission from the intelligence controlling the orbs or UFOs. Another example of the danger of UFOs being approached without warning is the famous Travis Walton case from 1975, in which he was hit by an electrical bolt of energy and was healed by extraterrestrial visitors who took him into their craft for several days.[83] JP says one needs to wait for the orb or UFO to approach them to be safe from such electrical discharges.

> JP - I saw liquid, a liquid type of substance on the floor. It looked like water, but when you kicked it around, it turned into a ball-like substance. And I think this was coming from the ships. The ships, they sweat this liquid that falls down on the ground. And you could just kick it around. This type of liquid is quite interesting. It looks like water. It acts like an oil type of substance. But when you kick it around, it doesn't stick to your feet, but it stays to the location where you kick it around, and it floats around... No, it's not mercury. It looked like water. Mercury is like a metal. This [substance, however,] you can still see through it.
>
> Yeah, they had a vacuum-looking machine with wheels. This machine had a big drum in the back, and it went from left to right, picking up this substance. And it was halfway full. I was trying to ask the guy what it was, but I saw it dripping from some of these ships to the floor. So it came from the ships. I don't know what substance it is. They recollect it, I guess. I don't know what they do with it. I don't know what the substance is, but yeah, I saw that. And it was quite interesting that we had a little machine with wheels picking it up, like vacuuming it up into a tank. So that was quite interesting. You know, I was wondering, why would they pick that up? And what is it for? What can they be using with that [liquid] that can help us, you know? So I saw that... While the pilots were leaving [on their missions], the hours were passing by. You had pilots coming back. You had pilots going. So, I was just

seeing ships coming and going. It looked like an airport type of training... It's quite interesting.

MS - Very interesting because I just did a quick check-up on Wikipedia, and this thing called hydronium, which is like water ... So I wonder if that's what it is. It's called H_3O^+.

JP - Okay, that makes sense. Yeah, I didn't know about that. That's really interesting.

MS - Yeah, it has properties similar to water, obviously, but it [behaves] very differently. Maybe you have to ask next time.

JP - Yeah, it was quite an interesting substance that I was looking at. Other than that, I didn't see anything else. I do have a missing time on that trip, a gap of two hours. That was when I was observing the orbs, and I was in one spot. And then I ended up in another spot without knowing how I got there. So, I didn't question that when it happened because I know I get downloads when I get home, and I know what happens after. So, I'll probably get a download of what happened during those two hours.

MS - Okay, actually ... hydronium has three hydrogen atoms and one oxygen atom, but still, it's based on water, just to be exact.

JP - Well, would it look or move similar to water, or does it look like water?

MS - Well, I just looked it up. Yeah, H_3O^+ is a kind of isotope of some special type of water. So, yeah.

The water-like substance dripping from the spacecraft is likely water (H_2O) that has gained an ionized proton (H^+) due to the powerful electrical currents ionizing hydrogen molecules coming into contact with the craft, which then combine with water in a chemical reaction to create Hydronium.[84] This is a possible explanation for the behavior of the water-like liquid that JP observed dripping from the craft.

> JP - And who's doing the research now, Doc, of that substance, you know?
>
> MS - It is actually a compound of water. It's called a protonation of water. I'm not a chemist, so I really am out of my depth here in describing what it is, but it's some kind of compound or isotope of water. Very interesting.
>
> JP – Interesting. So yeah, these ships are dripping that all over the floor, but it kind of moves like mercury, so it was quite interesting. On different types of rock, I saw a type of mushroom that was growing. It was like a blue type of mushroom that was growing on the side of the walls of this particular cavern. I don't know if that can ... [reveal] what type of cavern this is, or I guess the oxygen level. I think the oxygen level was really rich. I felt really awake in this cavern. I think the ship has a way of creating oxygen because that particular cavern ... where the ships were, I felt like ... I don't know if you've ever been to an oxygen bar. You get that same feeling in this particular room. So, I think these ships sweat oxygen and give out oxygen as well. So that is quite interesting.
>
> MS - And you had a missing time experience as well while you were there with the orbs.

JP - Yeah, I had missing time. So, going back home, there was a total of 18 of us. There were still people flying and doing, I guess, training or [on a] mission, whatever the hell they were doing. We drove back home to the spot where they picked us up. And [hovering] over us we saw a TR-3B flying really low. And I think one of the pilots who was coming back [with us], he was saying that he was going to fly that. I guess the ship knows where the pilot is in ... [real] time. I don't know; they're like pets. I don't know how to put this, Doc. It is insane how this technology is. And the pilot was saying, yeah, that's my TR-3B. And we saw it. It was flying over the bus to the location where they were going to drop us off. And I said, "Are you going to fly it now?" He's like, "Yeah, I'm just going in 15 minutes to meet up. And we're going to train in it." So, I guess, yeah, they train, and they fly. And yeah, these pilots are cool.

MS - That's interesting because I know with the Nordic ship or the ET ships that they have an organic consciousness that matches with the DNA of the crew of the pilots. So, it sounds like the TR-3Bs, or whatever this model is of this triangle-shaped craft that is maybe a more advanced version, that it includes this kind of organic [consciousness] that bonds with the pilot or crew.

JP - Yeah, the color, it wasn't completely black. I saw it was a grayish-blue color type of shit, but it's a similar size to a TR -3B triangular [craft]. It looks just like a TR-3B, but it's smoother, and more round. It's not rigid. But I can see the light still in the middle and the three lights in each corner. So, it was like a type of TR-3B, quite an interesting kind of ship. I could clearly see it because it was sending down a light, and it was making us notice that it was flying above us. I don't know if it's [giving us] a type of protection because after we parked and we got out of the bus, there were a couple of UH-60s [Black Hawk helicopters] that flew over us. Also, I think protecting the area.

Figure 26. A day-time photo from 2017 where JP previously saw a flying triangle & helicopter together

MS - Okay, so the bus arrived at the destination where you got on.

JP - When we arrived at the destination, there was a total of five of those [pilots], 18 had left. The bus was dropping people off in certain locations. I didn't know the location they were dropping them off ...

MS - But it ended up at the location, the base where you work, right?

JP - Close, yeah. So, I started going home, and a car followed me back to my house, making sure I was safe. And I think this was a big deal based on what I saw. That same time, I drove back to the base, and I met up with the person who gave me the green light [to talk]. He says, "Yeah, you have the green light to talk about that. Don't talk about the details that you saw on the walls. Just talk about how the United States has the capability of these ships. Yeah."

MS - So that's awesome, and that's worth emphasizing that you're not really a whistleblower because, you know, while you're active US Army and you're talking about these missions, you're being given the green light by a superior officer who's involved in the chain of command of these covert missions.

JP - Yeah, I even took you to the location where he talks to me.

MS - Yeah, that's right. I went on that base. You took my wife and I there, and you showed us that location, and yeah, hopefully, one day I can meet the officer. But right now, I mean, this all has to be confidential. But it does mean that there is a covert branch of the US military that is ...

JP - I could say that he does travel to Washington, D.C., a lot. I could say that. Yeah. I hope he doesn't get bothered by me saying that, but hundreds of people go to Washington, D.C., you know.

MS - Yes, well, it's good to know that there are people within the covert military that want this information to get out and people are always wanting for there to be disclosure. We all are impatient for it. But at the moment, it sounds like for senior officers to come out and openly reveal it, it would be too dangerous. But having ...

JP - Yeah, it is dangerous for them. You know, I guess what David Grusch did, you know, it's really hard to come out like that. I feel sorry for the guy. Yeah, I guess he took it upon himself to come out. And I think sooner or later, there's going to be much more people coming out with all this information. We've been saying that you know? All this has to come out.

David Grusch was a former USAF intelligence officer who appeared in a July 26, 2023, Congressional Subcommittee Hearing to share what he knew from up to 40 credible witnesses of classified programs involving the retrieval of crashed UFOs and the bodies of Non-Human Intelligence.[85] He further claimed that major aerospace programs were attempting to reverse engineer the recovered non-human technologies.

MS - And, of course, in the case of David Grusch, I mean, he really hasn't had the kind of access to these classified programs like you have. You've actually been on the missions.

JP - I'm sure he has, but he's really hesitant to talk about certain things that he did [or] saw and that he was involved in.

MS - Oh, were you talking to someone? Someone said that David Grusch was part of some of these missions.

JP - I haven't been told that, but I'm sure he has. He also has a green light and red light [communication] system that talks to him and tells him what to say and what not to say. Yeah, I'm sure, you know, to be in a position that he did and where he did it, I'm sure he has been through different experiences.

MS - Well, we're very glad that you come forward and talk about these experiences and that there are people in the military who are protecting you and making sure that everything is okay as you disclose this stuff. So, is there anything you want to say to our Portuguese and Spanish friends?

JP ... Yeah, I basically said that I'm glad that they're hearing this information on your channel, and there's going to be more things coming out. Let's see what happens on Earth, you know, on top.... I said imagine what the Inner Earth people think about us when they see us going to war and see us dying of hunger, see us living to 80 years old, you know, imagine what they feel when they see that from us. You think somebody would want to communicate with us if we're going through all these situations on earth, you know, black hats versus white hats, outer solar system, things happening to people, death, it's insane what's happening on Earth. I agree that certain civilizations are not communicating with us, and I understand their point of view. They help who they want to help, Doc—who wants to be helped and who wants to search for the truth. So, they are going to communicate with people who want this, not the people who are negative.

MS - Very true. So, I want to thank you, JP, for your service and your courage in coming forward and revealing all of this incredible information.

JP - No problem, Doc. It was a real pleasure talking to you. I love everybody.

Chapter 10

Testing for a Covert Space Mission to Saturn

On July 4, JP was sent to a medical facility at a military base where he currently serves, along with three other soldiers who had picked him up at a pre-arranged location. The soldiers were veterans of covert space missions and were dismissive of JP, who they didn't know as he hadn't completed the same training. They questioned why he was there and why they were sent to pick him up.

Once JP and the soldiers arrived at the medical facility, they were screened by a tabletop device that required placing one's hand on the surface that would determine their suitability for an upcoming space mission to Saturn. JP's handprint produced a green-yellow result, which showed his suitability and that he had previously performed covert space missions outside of our solar system.

The three soldiers' results indicated that they were less experienced than JP in deep space missions. The more aggressive among them apologized to JP for his disrespect and said he had no idea of what JP had previously done. JP was also very puzzled by the results as he doesn't remember doing covert space missions any further than Jupiter's moon Ganymede, which he traveled to in late 2021 as part of a space convoy headed by US Space Command.

When JP later met the senior officer, who was his handler, to ask about the upcoming Saturn mission, he answered in an enigmatic way that the mission may have already been completed. This stunned JP, who didn't remember completing such a mission. JP realized that he has completed far more space missions than he remembers and looks forward to regaining his memories. The transcript of the interview follows, with grammatical corrections and my commentary.[86]

Interview Transcript with Commentary

Key: MS – Michael Salla; JP – Pseudonym for US Army Soldier

MS - It is July 9, 2024, and I am with JP, and he has something to report regarding a recent incident that took place on a military facility. So welcome back, JP; it's been some time since we last did an update.

JP - Yeah, sure. I'm happy to be here, Doc. Thank you for the opportunity to bring this information. It's good for the public to know this information. What's happening on the inside? A lot of people need to know what's happening … Right now, we're living in the days when there's gonna be a lot of changes for a lot of people. For you and me, we need to bring more of this information [out] because there's going to be more people going into the military because of the draft they're doing. So, a lot more people are gonna experience a lot of different things in the military. I think they should know what they're gonna get into. So, I went to a parking lot at the military base. I was waiting there, and I parked my car in the particular spot where I meet up with my 'friend' that gives me the green light.

MS - And this is a location that you took me to when I went to visit you at that base.

JP took my wife, Angelika Whitecliff, and I on a tour of a very large military base where he was stationed in March 2023, and he described some of the locations relevant to his missions.[87] This included where he would meet the senior officer, who would either green-light or red-light what information he should release in his public mission reports.

JP - Yes, Doctor Salla, that's the location, and it's a little bit more down to where the parking lot is and where the boats are. I stopped there, and then a white minivan came up, and it parked. And then the guy [inside] said, "Are you so and so?" I said, "Yeah, I am that so and so." And he's like, "Okay, come on, hurry up, let's go, tap tap." So, I got in, and there were two other guys in the van. And I'm like, "Okay." There were two other guys here, and they were dressed in military uniform, but I was wearing civilian [clothes]. And the other two guys looked at me with a frown on their faces. I'm like, "Is there a problem?" And they're like, "Who are you? We've never seen you before?"

And I'm like, "I don't know, I'm here. They want me to be here." And then the other guy was talking to the other dude. There's three guys, and they're all talking to each other like, "Who's this guy that we're picking up, like a random dude, like we've been doing this for years." They were saying … "We've been doing this training for years, and now we got this guy like a random dude." And I'm like, "Hey, I don't know what the problem is. But I'm just here because they want me to be here, okay? So, I'm not gonna talk to you guys anymore." And then the third guy was more outgoing.

He's like, "Hey man, where are you from?" I was like, "Hey, I'm from New York, and he's like, "Have you been to space before?" And I'm like, "I can't talk about that now. I'm just here to do whatever they want me to do." So, he was trying to get out of me [information about] if I have ever been to space before. I guess the facility that … they're going to take me to is … like a training. But it was like the people that only go through these programs go through there. And it's for a mission to go to space. So, it's a checkup that people who go into space [have to] get to make sure their body is good and everything is in the right spot before going. It's a big checkup that they do. So, they took us.

We drove one mile into the base, you know, the base is humongous. So, we went to the interior of the base and to this facility. It looks like a random building. We entered this random building, and it looked like an office, and there were regular people there. And then they took us four to the back, to like a big metal heavy door. They opened it, and then we went in there. The door was at least a foot and a half thick of metal. It was huge. So, they took us to this room where the other people could not see us, and we went in.

And then they opened a metal door, and we went [through] that metal door, and they locked it. And then we saw a lot of doctors. There are a lot of doctors there walking around with face masks and gloves. I saw two other guys sitting down without [clothes on], only with their underwear sitting down ... They had a machine, like a square machine ... and this machine had a lot of cables coming out of it and condensed into one cable, and it was going into their veins. And he was like, [with] one arm down, and he was sitting down, and he had like a metal desk right in front of him. He had his other hand on that desk. This is what I did.

So, they sat me down, and I'm going to tell you how the procedure went. They sat me down, and they said, "Take off your clothes and stay just in your underwear." So, okay, I took off my clothes. I [kept] … just my underwear on. Everybody else took off their clothes, and then the other dude, who was really rude. He was like, "Man, why are you here, bro? Do you deserve to be here? What are you, like what's happening?"

And then the other dude that was more soft [polite], he was like, "Hey man, leave him alone. He's just here because they want him to be here." "But I don't think it's fair for the other guys that did not come. They're doing training, too. And they just told us that we were gonna pick up somebody, but I didn't know it was gonna be a random person. We had never even seen him in the training before. So why is he here?" And he was talking [aggressively] like that, and the doctor came in the room, and he was like, "Okay guys, I don't know what the discussion is here, but I think you guys need to lower your heart rate right now." The doctor came in, and he had a mask on. He had an operating room look overall, and he had the whole blue overalls. He had the gloves and he's like, "Hey, you guys need to sit down and get ready for this procedure." They sat everybody down, and they had everybody put their hand on this metal desk-looking thing. So, when you put the hand on it like this, numbers come up, and it tells you your heart rate and the sugar in your blood.

And it tells you some other code numbers that I did not know what it meant. And then, on top, it shows you a color. So, my color was greenish-yellow, and all their colors were bluish, like bluish-purple. I was looking at everybody's color. And then the guy that was arguing with me, he's like, "Holy shit, bro, look, he's like a greenish yellow color. That is effing rare, man." And after he saw my color, he started treating me better. I guess the color depends on how many missions you have been out on before. I think it's a lot of missions that I have been on before, and I haven't been remembering [them]... How many missions have I already done that I don't remember? So, it's possibly in the past or altogether.

Jean Charles Moyen also talked of placing his hand on a machine that emitted a blue or green light after evaluating key bodily functions to ascertain suitability for space travel. JP first discussed encountering such a machine in his documentary, Starseed 1, which he explained in a private communication as follows:

> The first times I talked about it were in several interviews on the Internet and in my documentary *Starseed 1*, when I explained what happened to me in the hospital basement when I was hospitalized for sunstroke at 13. The woman had asked me to put my hand in a small box to recognize my DNA in order to authenticate who I was, and a neon light, like a photocopier, lit up and ran across my hand. It was hot and the light changed from blue to green, and then she gave me the black suit with the Pegasus insignia on my chest. I had been recruited.[88]

Next, Moyen explained his reaction when he learned about JP's interaction with a machine at a military base that similarly assessed his suitability for space travel:

> Imagine my surprise when we were in France one night. JP and I went for a walk in the woods, and he told me the same details about the light colors. I knew he'd experienced the same thing as me

because I'd never specified those two colors. I'd only talked about a light that had traveled over my hand to identify my DNA.

My interview continued with JP continued as follows:

> MS - Now, I know you've reported on many missions to me over the last few years, and you said that often there was some kind of memory wipe attempt that they did, but you still remember the mission. But it seems that there were other missions that you didn't remember where this memory wipe did work.
> JP - Yeah, but sooner or later, the memories come back. I don't know how, but the memories do come back. And it was when he started treating me [that I] ... I got scared. I'm like, "Man, I don't remember doing as many missions, you know," And I guess the color also means how far you've been from Earth as well. So green is really far, greenish yellow. So, he [the aggressive guy] was like purple blues, the farthest [for that], I think, was like Mars or Jupiter. But when it's green and yellow, you've been further out into [or outside] our solar system, and I found that, like "wow', you know. I kind of got scared because the missions I remembered were in our solar system. That's not unless, you know, the domes were somewhere else from our solar system. So, I'm like, "Okay, of the ones I remember, you know, the farthest I went out was like Jupiter, to Ganymede.

JP has been taken to the region of Jupiter by both the Nordics and US Space Command on separate occasions. Space Command led an international convoy in November 2021 to Ganymede to meet with a new group of highly evolved extraterrestrial visitors arriving in the vicinity of Jupiter. JP spent time on Ganymede visiting ancient extraterrestrial facilities there.[89] JP also has been taken by Nordics to resupply extraterrestrial bases on both Ganymede and Europa in November 2022.[90]

> MS - OK, so there was something in your blood that indicated that you'd been on missions outside of our solar system, and they could detect that.
>
> JP - Well, the way they were explaining it is that our body is made up of 75% of water or something like that.... A big percentage is made of water. Soon, we'll get that information out. But your body carries that memory, a memory of how many places you've been out to. So, the liquid of your body or your blood carries the memory of how many times you've been out [into space]. I guess blood [also] carries memory from your ancestors and all that.
>
> So, they had me put my hands on the machine, and then it turned green with a lot of greenish, yellow, a lot of numbers. And they had this other hand [needle] go down, and they sticked it into me. They sticked a needle [into me], and my blood started coming out. It did hurt. The doctor said that they actually put a numbing lotion on first, and then they sticked me with the needle. The needle is quite thick. So, it went into a machine, a portion of my blood goes into there, and it comes back out with cleanliness [purified]. I think it's similar to what people with acute kidney problems do, dialysis or something like that.... I was just sitting up, and I was feeling dizzy and my hand was down.
>
> The doctor kept coming in, "Are you guys okay? Are you guys okay?" And the other dude was almost crying, saying, "Oh, I can't believe he's a green,

greenish-yellow." And then he apologized to me, saying, "Hey, I'm sorry, man. I didn't know who you were." I'm like, "No problem, man. I don't have a problem with you. You know, I'm sorry you feel the way you feel, but I was told to come here." And he was like, "Are you allowed to talk [about] the places where you've been?"

So, I told him, "Yeah, I do talk about the places where I've been, you know," but I never told him who I was…. A lot of them know, … they listened to you, and they're trying to find out. "So, who's JP?" You know, because they want, you know, they want to say, "Oh, I met JP, and sometimes I don't like to talk because my voice is really unique and people recognize my voice. So, the second dude, he was laughing because he knew who I was. He knew my voice, okay. So the third dude was more soft [polite]. The end guy, he was like, "Man, I didn't realize you were a green. And bro, I'm really happy that I got to meet somebody like you."

MS - Do you know anything about the other colors? What were the colors of those guys?

JP - They were bluish-purple.

MS - And just explain again, because I must have missed it, bluish purple, exactly what were the different colors' significance?

JP – The colors mean how far you have been from Earth; that's how they explained it.

MS - Where do they extract that?

JP - When they put your hand on the table, all that information comes up.

MS - I see, so it's just a kind of hand technology machine.

JP - ... When you touch the metal table, it gives you all that information. And you feel like your hand got stuck to it... It's like a magnet, feels like a magnet. It feels like the bottom of your hand has a magnet. And when you touch it, it doesn't come out easily. So, you're like, "Okay, I guess they locked it." Like they have a mechanism that locks it. The doctor has the mechanism that they lock your hand [in position] there. So, it's an interesting medical table. I haven't seen a table like that. I don't know what technology this is or where this technology is from. I know the Nordics have a similar technology on their ship. Like they put their hands on certain things and it makes other things move. But I didn't know where this type of table came from.

JP - So after they were taking out my blood and all that, they sat us up, and they put us in a booth. So, we were sitting up in a booth ... our clothes were off, and we were getting like a shower of, it looked like particles of gold, and our whole body turned into like a goldish color. You can see our whole body [covered] with a golden color and it looked like I had sprayed on sun lotion. So everybody was looking golden. And I looked at the guys, and they looked at me, and we all looked like gold dust from that wrestler, Golddust. I don't know if you've ever heard of them from WWF.

> Everybody was golden [looking], and then we came out, and they sprayed us down. They took all the particles out. And that procedure, I don't know why they did that procedure, maybe to figure out the body type that you are. But the doctor did not talk a lot about the procedure. The only ... procedure he talked about was [about] a table and the color formations and all that. After that, they told us to put on our clothes, and we were in line, and we started going down this ramp and ... the three guys and me, they sat us down. It was similar to where I went with Dan. You remember when I went to a place with Dan that there were other ETs there, Nordics and other people there as well, other military people.

JP is here talking about the mission discussed in Chapter 7, where he and Dan were taken to an underground castle in an icy region of the US where shipping containers were being loaded with gold to be shipped to international locations.

> JP - So, some people were dressed randomly, and some people were dressed in military uniform. We were all in line. And then there was a dude in the front who was like, "You guys know why you're here." And I'm like, "Oh man, here we go again. I don't know what the hell I'm doing here." And I'm just nodding my head and saying, "Yeah, okay," to whatever he says, "You guys know why you're here, and there's a mission going on." So right there, I realized that I was gonna go on a mission, and they were preparing everybody for a mission, for a certain mission.
>
> So, I was thinking to myself, "This could be one of my last missions." And I was thinking like, "Where

would this mission be to?" Like, "What's gonna happen in this mission?" So, we started talking to each other. We're like, "Hey, where do you think we're gonna go?" And they're talking about the rings of Saturn. And I was thinking, "Is there gonna be a mission to Saturn, and what are we gonna be doing there?" So he was telling me that there were ships that were coming and then that we're gonna do rendezvous with these ships that were coming in.

But he didn't tell us who or what type of ships were these that they were gonna be coming. After the briefing, he told us more details, but I didn't get the green light to tell more details about what they told us in that room about what type of mission it was and what we were gonna do. But I could tell you that we're gonna do a rendezvous with another particular ship. And I know what we're gonna be doing, and I know what type of mission it's gonna be, but I just can't talk about it now. But I could tell you that, yes, it's gonna be to Saturn, and it's gonna be a connection with another species, another alien ship. We're gonna connect with them.

MS - I see, so this testing was to really assess whether or not you could do that mission safely?

JP - Yes, because it was farther from Jupiter. So, because I had a green, greenish-yellow color, they said, "Yeah, you're going on this trip." Interesting, so not a lot of people got that greenish-yellow color. We got out from there, and we went out. We went back to the metal doors, and they closed. And the guy who was cursing me out, he put his hand on my shoulder and said, "Hey man, if you need me, I'll be here for you. You know?" I was like, "I'm all right, man. Thank you, brother. It is what it is." So, they took us back to where my car was.

I sat in my car, and then the person who gives me the green light, he came up to me about 10 minutes later and he met up with me. So, he knew I got dropped off. He's like, "So, what do you think? Are you going? Do you want to do this?" And I'm like, "It's a dangerous mission. That's the thing." So, he really was asking me if I really wanted to do it. It's a dangerous mission. And I said, "Yeah, I will do it. When is the date? When is it?" And he's like, "It's not the date. It's when you did it."

MS - Oh, wow.

JP – So, I looked at him, and he started laughing. "Welcome back," he tells me, "Welcome back!" So I'm like, "Wait, wait." And he's like, "No more, you got the green light to talk about this, but then you'll start getting... We haven't taken everything away. You're going to start getting feedback on what you're going to remember about this mission.

MS - Okay, so you have done a mission to the Rings of Saturn, but you are not aware of it. The memories are not there.

JP - So, it's a type of time or separation of consciousness technology that takes you to somewhere else, to another body, or to another, maybe, a clone of yourself, or a place that is different. So, yeah, that was it, that's what happened to me. It was quite weird. When did this happen? This happened. I think I texted you. I'll tell you the right date, the 4th of July [2024].

MS - Okay, so it's the 4th of July that you went to this facility?

JP - It's more of in the morning... Yeah.

MS - So on the 4th of July, you went to this facility and underwent that kind of testing, that experiment. And you find out that this is kind of some kind of screening, some kind of process to see who's suitable for doing missions. You're capable of doing missions, or you've done missions outside of our solar system. So, I guess it's not a surprise then that you've done one further than you previously remember. That you remember going again and again, [Ganymede] is your furthest mission as far as you know. But now you know that you've done one to the rings of Saturn, that it was a dangerous mission. But you don't remember any details about what happened at that mission at this point.

JP - I don't remember as much, but I know something is going to come out from this that is really interesting.

MS - Do you think, based on the feedback you were getting from those other soldiers when they're looking at the green-yellow reading from your handprint and that suggested that you've done missions out of our solar system, that there's a lot more memories?

JP - I think there are a lot more memories. Yeah, that I don't remember.

TESTING FOR A COVERT SPACE MISSION TO SATURN

MS - Interesting because we've done, I think, 35 mission reports so far. Since you've been telling me about your different covert missions to various places, 34 mission reports, so, there might be another 34, or there might be much more. We have no way of knowing exactly how many more missions you've done.

JP - I've been in the military for almost four years, so it could be way more.

MS - Right, and of course, the missions that you remember, I mean, these might actually just be the surface level of what you've been doing, that the other memories may be, I don't know if it's due to trauma or dangers.

JP - I think when it's dangerous or when you lose somebody, or it affects you, you get hurt in it, or you see something that you weren't supposed to see, that's when they start taking the memories away. But I don't understand, but I think I've seen it all, you know? Well, I think there's certain things that I'm not supposed to remember. I think it could be because maybe it's in a different time, and it could affect our future if I do know that particular mission, or maybe it happened in the past or in the future, and I can't talk about it because it could affect the outcome of a lot of things on Earth.

JP raises many scenarios here for why he may have had his memories of some missions removed for a variety of reasons. This suggests that in the future, when conditions change or for other reasons, JP will remember many more missions he performed during his Army career.

MS - So, your superior in the covert hierarchy told you it's not a question of when you're going to do the mission. It's like when you did it. He gave you no clue as to when that happened a week ago or a week previously. No clue. So, do you have any idea?

JP - I was out for four hours in that facility. It was the early hours. I got there at 9, 10, 11, 12, 1300, [and by] 1400 [2 pm], I was heading back home.

MS – So, do you have any idea when the memory of that or the other missions is going to start bubbling up to the surface? It's guesswork at this time.

JP - No.

MS - Well, that's a fascinating experience, and any of those guys that ...

JP - They did not look familiar; I'd never seen them before. The guy that probably knew who I was, he was really quiet. Very discrete, but they were all, "Man, there's a lot of us that go through these experiences."

MS – Okay, but very few actually talk about them publicly.

JP - They don't talk about it publicly or are not told, and I actually asked the guy who gives me the green light, "Say, why is it just me?" He said, "There's other people reporting, but they're just reporting to ... other people that are in Washington [DC]. They're reporting to other people like governors who know that. So, you have people who are reporting to certain [high-level] people, but we'd like you to report this to the public. So, what we're doing here is a big key. I think we're opening the doors for these other people that are, I think, gonna start coming out. And they have been coming out like with that general that just came out saying that it is real that we have biological entities

MS - Oh, you mean that Admiral Tim Gallaudet?

JP - Yeah, that one. And I think they feel more, you know, free to talk about it when they hear my mission. So, like, "Oh, look at this guy, you know, he's visiting the Ant King. And you know, we can't talk about our little missions that we have done in South America or around the United States." I'm sure there's a lot more people doing a lot. It's a network, it's a job, you're told what to do.

Rear Admiral Tim Gallaudet, a former head of the National Oceanic and Atmospheric Administration, has made some startling public claims regarding the existence of Non-Human Intelligence and that official contact has been made.[91] It's possible that he is being secretly briefed on some of the covert missions that JP and others have experienced concerning Non-Human Intelligence.

MS - Right, with the Ant People, I remember the last mission you did; you said that you thought you were going to be sent back to find out what was happening in that other realm where the Ant People went and where the new king had taken them, that you felt that they were, like, just preparing you and you said, "no, I'm not ready then." But do you have any idea whether that's going to happen?

JP - Or it's already happened.

MS - Or it's already happened. Okay, but you just don't remember.

JP - Remember, it's a different realm. So that means it's a different time. So, I don't know if it's a time that jumps into the future or a time that jumps into another realm. So maybe they don't want me to remember [so as] to not affect this realm and this timeline....

MS - OK, well, wonderful to have you back again, JP. I look forward to further reports. Let's see what happens. If you do remember some of these past missions, we might be doing a lot more of this.

JP – Yeah, I want people to understand ... not to be scared, you know. I know I'm talking about these certain missions, and I know there's a lot of people that don't believe, you know, because they they want proof. But I have been releasing photos, I have been releasing videos, and you can search them up also on the [dedicated JP] page on ExoPolitics.org.[92] If I do come out with these other videos that I have, it's gonna be really exciting to show these videos and pictures that I have.... I have to be careful now because of my situation, as you know.

MS - Well, I look forward to seeing some of those other pictures and videos, and hopefully, we will find a way to release them without putting you under any kind of threat. So, I want to thank you, JP, for being on Exopolitics Today and this new update. JP - No problem, Doc. Appreciate you.

While JP did not remember any details about his apparently completed mission to Saturn, I was able to get

JP to do a roundtable discussion with Elena Danaan to discuss Saturn and its significance at a conference we all attended in Valence, France, in July 2024. We also addressed many issues raised in some of the other missions discussed in earlier chapters.

US ARMY INSIDER MISSIONS 3

Chapter 11

Roundtable on Space Arks, Sleeping Giants, ET Assimilation & Mysteries of Saturn

In this roundtable discussion, JP, Elena Danaan, and Dr. Michael Salla discuss multiple missions recently completed by JP, who at the time was still actively serving with the US Army, and how his missions relate to information shared by the Galactic Federation of Worlds. The missions have been discussed in earlier chapters and concern the relocation of the Atlantic Space Ark by Nordic-looking extraterrestrials, the use of crystal jewels hosting the organic consciousness of the Atlantic Ark and similar large spacecraft, the awakening of sleeping giants in stasis chambers connected to legends of missing Anunnaki scientists, the assimilation of Nordic extraterrestrials into human society, and a dangerous mission to Saturn which JP was tested for but does not recall completing.

Critical aspects of space and interdimensional travel in terms of its toll on the human body are discussed, as well as how memories of additional JP missions have been removed for safety reasons.[93] The transcript of the roundtable discussion follows, with grammatical corrections and my commentary.[94]

Interview Transcript with Commentary

Key: MS – Michael Salla; JP – Pseudonym for US Army Soldier, ED – Elena Danaan

> MS - I'm with JP and Elena Denan at a conference in France in Valence, and it is July 10, 2024. We are going to be discussing some of JP's missions and

what Elena has learned from the Galactic Federation and the Seeders about some of the things that JP has encountered during his missions. So, I want to welcome JP and Elena to Exopolitics Today.

ED - Thank you, Michael, for having us.

JP - Thank you, Doc.

MS - Well, let's begin with the Atlantic Space Ark.... You've been on six missions that you recall to the Atlantic Space Ark. On one of the more recent missions, you described both going into the space ark with a team and extracting a crystal jewel from somewhere and that it was taken away and used to activate technologies and arks elsewhere around the planet. And then it was returned, and the Nordic extraterrestrials played a critical role in that, and you were part of the mission to return that crystal jewel to the Atlantic Space Ark. So, you want to kind of add to that, elaborate on anything about that mission?

JP - It was a beautiful mission. The feelings I felt when we grabbed the jewel, it was emotional. You felt what the ark felt. The jewel is connected with the ark. So, when you hold it, you feel a sensation of sadness and happiness. It's a mixture of feelings that you feel. Maybe also, maybe, feelings that we never experienced before. I'm sure other ETs have different types of feelings that we haven't experienced before or felt before. So, it's a mixture of feelings, sadness, happiness, and some other feelings that I never felt before, which just make you cry and make you happy. Sometimes, it puts you in a position where when you blink your eyes, you see flashbacks of how the ark was. I'm sure that there are different types of jewels all over the place of these types of jewels that I'm sure hold the consciousness or the energy of the ark. And I think that's why they wanted to extract that particular jewel to feel what the Ark was really feeling.

MS - I remember you saying something like that it was somehow attuned to the Earth's vibrations as well and that you guys went through these feelings of joy and sadness. It was like you could feel the full extreme range of emotions.

JP - We could feel the different emotions of the ark, yes. The frequency of how the ark feels, where it is, and who's been in control of the ark. So, it's like a type of memory or a frequency that captures time and space and makes you feel like that, makes you feel all those feelings of, I guess, past civilizations that have been through the arks.

MS - And do you recall anything? What can you recall about that jewel being taken to activate other technologies or other arks? What do you recall about that?

JP - We just picked up the jewel, and from there on, I don't know what they did with the jewel.

MS - Okay, so you were just guessing, speculating that they used it to activate other technologies, but you weren't briefed, you don't know for sure.

JP - They said something about different technologies, but I'm not sure which type of technology they're using that for. I don't know if it's to activate other types of vehicles or arks that were dormant. I don't know if the other arks even have that type of jewel. But I know for sure that the Atlantic Ark had that jewel. So maybe they were taking that jewel out and probably putting it in another ark just to see how the other ark reacted to it. Remember, all the arks are cut [and pasted] in space as if they were built somewhere in space, and they're alike. All the rooms are in the same spot as all the other arks. So, when scientists or archaeologists do research on other arks, they phone or communicate with other people who are on the other arks. Hey, how about this room over here? Do you have this and that in that room? And we communicated back, saying, yes, we have this and that in that room. And it's similar. Yes. So, we investigate more particular arks. But they're all connected.

MS – Great, okay, so Elena, what do you know of the Atlantic Space Ark and this jewel that seems to be a critical part of its operations?

ED - What I know about this Atlantic Space Ark is that it was the base of the construction and development of the capital city of Atla, the capital city of Atlantis. Atlantis had two capital cities, one that was older, on the bigger island, that was the headquarters of Enki, or Ea, when he used to live here for a period of time. He gave territories, especially islands in the southwest, to the Alteans, who were part of the Seeders, the Intergalactic Confederation, to come and set up a colony there.

So, they brought that spaceship, what we call the ark, there. And what a lot of ETs are doing, this is very customary for them to do this, they will be able to build their colony and their infrastructure around the arks because the arks have all the survival material for them, you know, all the technology, the medical, everything, and the energy supply. So that's what they did with Atla. So, this ark was the core center. And when the big Younger Dryas event occurred, the whole Atlantic saw the sea rising, and Atlantis was destroyed, of course, and all the islands [sank], the ark sank and went deep down. All the infrastructure that was built afterward was destroyed, but the ark sank and stayed there.

The jewel, I suppose, JP is talking about what I, 100%, am sure, in fact, it's not supposition—it's the core consciousness of the ship. So, these ships are half organic, and they have a consciousness that could be interpreted as Artificial Intelligence, such as in the ships of the Galactic Federation of Worlds. The scout ships I've been in for the Galactic Federation of Worlds have Artificial Intelligence, meaning a super-developed computer program that is able to think for itself. That's not the case with the Seeders [ships]. It's not because they are 20,000 years in advance of the Federation. They embed real consciousness, plasmic consciousness, into technology because it's way, way, way more advanced.

So, these real consciousnesses are beings. They are out of time. They don't have the same notion of time as us. They will be embedded. They will accept being embedded in a ship, and the ship will become their body for a while. The way to contain consciousness is in crystals; it's the best way. Consciousness can be put into crystals. You know, I can mention the crystal skulls that can hold for a while, receive for a while, human consciousness. It's the same for some types of crystals in a spaceship. So, I suppose that what JP is mentioning is that it was the consciousness of the ship. Removing this consciousness from the ship means you cannot work the technology anymore. The ship is dead. No matter how you try, no matter if you have the right DNA, it won't switch on. So, it's a way of protecting the technology that is hidden in the ark to take the core consciousness away so then you cannot use the ark. I do not have this information about where it was taken either, but it is probably in a safe place.

MS - So for them to remove that crystal jewel, containing the consciousness or the organic consciousness of the ark and taking it away for a limited period of time, what would they use that for? I mean, JP mentioned activating other technologies. Do you have anything that comes to mind in terms of the practical use of this kind of organic consciousness stored in the crystal that could be used for things other than animating a ship?

ED - We have two options here. Either this was taken in a safer place by the Intergalactic Confederation, which would make sense to me. Or was this used by not-so-friendly intentions to animate other technologies? Maybe some bad-intentioned people have exotic technology of the same kind in their hands, and they need a consciousness to inhabit this technology to make it work. Maybe they could try. But I'm quite pessimistic about the results because these are organic beings, consciousness. You need them to agree to that. When we embed a

> consciousness into a ship, if it is an organic being, the being needs to agree and be okay with that. If the being is not okay with that, nothing will work either. So, you know, I don't know more.
>
> MS - What comes to mind is David Adair's experience with this engine at Area 51 back in 1971. He says when he interacted with this engine from an extraterrestrial, from a larger extraterrestrial spacecraft, the consciousness, he communicated with the consciousness, and then it transferred into him. And he says that it used him as a lifeboat. So, I assume that this organic consciousness, which he called his Pitholem ... can be transferred either into a crystal or even into a human body temporarily.

David Adair has revealed in many interviews how he interacted with the consciousness of an ancient space engine stored at an underground facility at Area 51. The engine was alive when he first encountered the consciousness, which told him that its name was Pitholem and that it crashed on Earth millions of years ago during a space battle.[95] When the consciousness downloaded itself into Adair, the extraterrestrial engine went dead, just as Danaan described earlier. Importantly, Adair added that a fleet of spacecraft similar to the one Pitholem previously animated would return to Earth 50 years later. This matched Danaan's information about the return of the Seeders.

> ED - Yes, if it is transferred into a human body temporarily, the human body needs to have the right DNA matching the frequency of that being. So, as it is from the Intergalactic Confederation, the Seeders group, it needs to have some percentage of this DNA, of a certain type of DNA, you know.

MS – Okay, so David Adair had the right DNA for this consciousness to transfer into him.

ED - Certainly.

MS - Great. So, on the return of that crystal jewel,

JP, you said that the Nordics ... had possession of that crystal jewel and that they returned it and that you were part of a team of four soldiers and a team of four Nordics that returned the jewel, the Atlantic Space Ark jewel, to where it was originally taken. And again, you went through the emotions, and you described the Nordics also getting teary-eyed and also going through the same emotion. So, anything you want to add to that, the return of the jewel?

JP - When we saw the Nordics having emotions, you know, they're like Vulcans. They don't have as much emotions. So when you see them having emotions, you know, it's something big, you know, it's something powerful because I think they control the emotions just like the Vulcans in Star Trek. It's kind of cool how when you see Star Trek, it's like a cut and paste of what's happening in the real world. So, we felt the same feeling in the crystal; the crystal actually grew brighter, like it felt happy to be back, and it was glowing. That purplish fluorescent light, beautiful, beautiful purplish with the lightning stripes inside of it, was glowing.... It felt a more beautiful feeling when it was coming back to the ship. And like Elena was saying, the ship was activating as well, certain parts when the crystal was coming back.

So, when we put it back in place, everything was activated on the ship. And that's when the Nordics started going to different parts of the ship and activating the ship as well. So, we just left it at that, and we left the Nordics in charge of the space ark. Because, you know, I said this before, like if you leave a cell phone in, in the middle of the Amazon with these indigenous people, they won't know how to play with the cell phone. Now, unless somebody shows them, I think it's the same thing.

The Nordics know more and have more knowledge of [the ark] technology than our military does. You know, but we're learning as much as they are, but they have more knowledge and technology. You have to be frank with that. We left them with the ark, and they are activating it. They were activating the ark in different ways that we didn't know how. So, it was quite amazing seeing that. And I think it's like that. I think the jewel could be like the heart or the brain of the ark. You know, as human beings, we could do a heart transplant, you know, move a heart to another person, and the other person can live with the other person's heart. I think the jewel has that capability of activating other stuff as well, or other arks because it's like a pineal gland of the ship or the main consciousness of the ship that could be like a heart.

Like when we do a human [heart] transplant, you can move it to another body, and the other body activates and lives again, because it has another [heart]. But knowing that the heart also has 40,000 neurons that can also connect with the brain and share that same knowledge of where it was before. So, it's quite interesting.

MS - Elena, do you want to elaborate?

ED - May I comment?

MS - Please.

ED - So JP, the jewel that was taken from the ark, was it taken to another ark?

JP - I think they were taking it to other places to activate.

ED - Okay, so that changes what I was thinking. So they were, I think, the people, these Nordics, as you described them, they were Alteans. It explains the fact that they were emotional because they were the original race that built Atla, who was linked to this ark. And there are, of course, other arks, as you describe, in the Atlantic or wherever else. And they used this consciousness to temporarily activate other arks. So, they weren't taking this tool to another place to be safe, or it didn't fall into the wrong hands at all, to try to activate some technology, as I thought at first, as I said earlier. So yeah, it's all good. So, these Nordics are the original part of the original Atlanteans. And so yeah, let's understand that.

JP has never asked about or described the origin of the Nordics he has encountered during his contact experiences, beginning with Brazil in 2008 or with the Nordics working with the USAF, which he has encountered on several occasions, such as Orlando in 2018. When asked where they might be from, JP said he had the feeling that the ones that first contacted him were from the Pleiades. This has some significance for the Atlantic space ark, as we will see shortly.

> MS - And maybe it's good to just clarify because you've said that the Seeders are 24 civilizations from other galaxies, but then they have seeding projects all over our galaxy and Altea, which is one of the Seeder groups, but that they also work with other extraterrestrial groups. And so this group that gained possession could have been maybe an Altean colony in, say, the Pleiades or something like that. So, they could have identified themselves as Pleiades, right? And I think that it's important that the Seeders are working with extraterrestrial civilizations from our galaxy. So there are 24 Seeder races, and each of those races might be working with, what, a dozen or maybe 50 or whatever civilizations in our entire galaxy.
>
> ED - Yes, exactly.
>
> MS - Okay, that's important. So, JP, you described the Nordics taking control of the space arks, and you also mentioned how the space ark was relocated physically from somewhere in the Bermuda Triangle region to the mid-Atlantic region. As I recall, you explained at the time that this was because the Nordics felt that the US influence over the space ark was becoming a problem. They [the Nordics] wanted to make it more of an international cooperative venture, to move it into or towards the mid-Atlantic so that it would be more of an international effort as opposed to just a US-dominated effort. So yeah, you want to describe that and, of course, the Nordics' role in doing that.

291

JP - Well, it wasn't necessarily like a problem. I think it was the US's choice to let that happen and share information. Other countries have different arks, and they also have information from their scientists, archaeologists, and people. So, I think it was a mutual agreement with the Nordics and the US to the ark to move more into the middle of the Atlantic. And I'm sure on the way over here, Russia is bringing their armada. They probably pass through there and get information, and they're going to go back to the same route through the Atlantic and also grab more information. I'm sure all that is happening while we speak.

I think it was a mutual agreement. It wasn't them [the Nordics] trying to take away from the United States. It was a mutual agreement with the Nordics to bring it more into the middle [of the Atlantic] to communicate with all the other countries ... I think India was involved—something to do about Vimanas, activating Vimanas. They found a couple of Vimanas in India, you know, because the [Vedic] Indian Empire was really big back in the days, and it covered Afghanistan and covered Iraq and all that. So, I'm sure there were Vimanas in the bottom part of the sand that still needed to be activated.

You guys know what Vimanas are, and also, they [Nordics] are communicating with the giants of Kandahar and, I don't know, other civilizations in the Inner Earth. So, yeah, they decided to move it more into the middle and, I'm sure, closer to the [submerged capital of] Atlantis. I guess the ships activate better when they're in their original spot and communicate better with the frequency of Earth and the frequency of other races...

MS – So Elena, do you want to say anything about that?

ED - Well, I agree with JP on the fact that there is an international coalition, which is part of the Earth Alliance, working with the Seeders, the Intergalactic Confederation. The Nordics that you describe are part of them. And I think these ships that we see, military ships from Russia, that's it, coming on the path of these arks, are part of an exchange plan for maybe technology, you know, because the Earth Alliance is aware of everything about these arks and the technology. So, I would agree that this is not negative at all to see all these warships in these areas. And I like the fact that you have, you know, someone who can trace the roots of these ships and match the location of the arks that JP describes. So, I think it's a positive thing. Good confirmation.

MS - Yes, you're referring to Ruezo Zanrico. That's a pseudonym. He's someone who's working for a military contractor, so he can't reveal his identity, but he's been tracking ship movements in the approximate area where JP identified the Ark. He has found ships behaving in a very strange way, just stopping and in an area where there shouldn't be any reason for ships to just be stationary. And he's talking about a number of large ships. So that's just corroborating what JP is saying, that the space Ark is somewhere in a particular area of the Atlantic, and there's a lot of naval activity around it.

I discussed Ruezo Zanrico in Chapter 4, where he tracked anomalous ship activity in the areas of the Atlantic where JP described the Atlantic Space Ark was located both prior to and after its movement. In both locations, Zanrico found anomalous ship activity that could be naval vessels from different nations congregating around the Atlantic Space Ark.

> MS - I want to move on to another mission that you [JP] did, and that concerned Nordic assimilation, where you were part of a group of four soldiers, and you were met by four extraterrestrials who had arrived. And you described four of them as each looking different [to normal humans and] looking like extraterrestrials and speaking in a way similar to extraterrestrials. And then you took them to a facility on this large military base, and they went into that facility, and they came out with completely [different looks]... Their hair was cut short. They had normal clothes, and they spoke differently, and they had passports and identity papers. So, do you want to just elaborate on that mission?
>
> JP - On that mission, I was surprised when I saw the Nordics, right? We took them to this facility, and then they came out all changed. Right there, then, and now, I could see that they could be among us. They can be disguised and be walking anywhere. And they can be different in every place that you look, you know, we could be looking at a Nordic right now somewhere in the [2024 Valence, France Conference] event, and we won't know it because their facial structure changes to our structure.
>
> The Nordics are more like square and more, you know. You can see, you could tell the structure of the face when you see them, that they're not from here. Their eyes are different. So when they came out looking different, we were like, "Hey, that's crazy." Like, "Wow, I can't, I can't believe that!" And

I'm sure it's not just the Nordics; it's other ET races that are probably walking among us. And they, they learn the language, they're so fast at learning the language and the accent. They pick it up like really fast. Remember, they activate more parts of the brain than we activate. So ... they're like sponges. They can capture [information] really fast. What, we use 20% of our brain.

ED - Yeah, not even.

JP - Like that, they probably use about 80%, and they know how to use 80% of their brain, so they capture accents easily. They capture movements, gestures of what race they choose, and it's amazing. It's amazing when you see that. I think they even put a little bit of fillers or Botox or something to make [them] look more like us. And they're like regular people, they come out [of the facility] like regular people talking like us, you know, outgoing [extroverts].

They're a more communicative type of people. They like to connect with other people, and that's what type of person they are; [they have] outgoing personalities. They like to ask questions, and they like to talk to you. They like to learn about different types of technology, and they pick up on computer stuff really fast. They pick up on different types of technology really fast. They know how to do coding. They're really good with numbers and music because music is considered a language as well. So, they pick it up so people use the same part of the brain for music as for language. So I guess they know how to do music as well. It's amazing when you see that. For me, it was like, "Wow, we do have to be, you know, vigilant because if they could do that, the black hats can do that as well. So, we're like, "Wow, we don't know who's walking among us!"

So I went to New York, and I went to Times Square. I was looking at so many people walking around, and I thought to myself, "Wow, how many ETs are around, walking among us right now in Times Square? You see so many faces, so many different types of people in New York or in a populated area. You never know who's around and who can approach you.

MS - You mentioned identity papers as well.

JP - Yeah, they have all that identity paper. They come out with passports. They come out with IDs. It's like everything is done there in that particular building. I guess they do it so fast, like it was fast the way they did it. It's kind of like, wow. They go by the departments [in the facility] really fast. And they come out ready to go.

MS - So this is like an official assimilation program, just like it was depicted in the Men in Black movie.

JP - Yeah, yeah. Exactly.

MS - So, Elena, what do you know about this? I mean, is the Galactic Federation involved in something like this, or are the other groups doing this?

ED - Yes, that's the Galactic Federation of Worlds. That's the Earth Alliance program of assimilating Galactic Federation of Worlds personnel. These ETs that you describe, I was listening to you, and I thought JP was describing exactly Thor Han's human race, the physical characteristics, the character, the personality. You described totally my friend, Celadion, who is a pilot in the Galactic Federation of Worlds and who has long hair and high cheekbones, and he has this personality of always being curious and sociable and interested.

It's them, the Ahel [Pleaidians], and the eyes are different as you describe. They are wider, and they have a kind of blue metallic light in them. There's this program of revamping [physical alteration]. They do that with not all the personnel of the Federation entering the Earth Alliance work as infiltrated agents in society because Thor Han really disclosed and confirmed recently that there is a program of infiltration that is very big, very big. They don't do that with all the personnel because some personnel, such as the Alpha Centaurians, don't need revamping. They look like [normal] humans, and you can't tell the difference. But some human races, such as indeed the Ahel Pleiadians, need revamping because you see that [they're different]. They have high cheekbones and a very square bone structure, and you can tell they're different. So, it all matches, yes.

MS - I first came across this when reading Howard Menger's contact stories from the 1950s. He described how he helped these extraterrestrials assimilate that they would come in with long hair, and he would cut the hair. He would help them learn the American colloquialisms so they could fit in ... He would help them. But, of course, that was all done informally. And that was in the 1950s. Since that time, the US government has developed an official program to supplement or take over that spontaneous ad hoc assimilation [of extraterrestrials], which is interesting.

JP - I also want people to understand that it's not just males. It's males and females doing this [assimilation]. You know, I had somebody asking me, "Oh, you only talk about males." But when I talk to Nordics, I'm talking about males and females that are going through these processes and in these types of missions.

ED - I also want to comment on the languages. They have a way of learning languages by downloading that language in some parts of their brain. That's why they're so fast.

MS - So, you [JP] in 2018, you described an incident where you were living in Orlando, and a flying saucer landed, and a guy, a Nordic, came out with crew-cut hair ... He was wearing a distinctive Air Force uniform.

JP - Jumpsuit.

MS - Jumpsuit. But it was a patch. You said it was a patch that was unique to them. So, it seemed that within the US Air Force, there was an elite squadron or group that were these assimilated Nordics that were working with the Air Force, and they kind of assimilated ... but they had their own patch so that they could be identified from other Air Force squadrons. Do you remember that? Do you want to describe that?

> JP - Yeah, when he wanted me to go into the ship, and I decided not to, I wish I would have gone into the ship and explored it, but I just didn't feel right because I knew there was somebody looking at me. There's somebody making sure you know. I didn't want anybody to get hurt, so I was being foul (rude), so I did not want to do that. But yeah, he had a crew cut. He looked like a regular military [guy]. He was a little bit more pale, you know, but his his eyes were really bright blue, like a greenish-sea blue. He was inviting me into the ship, and I decided not to go. I felt so bad because I knew it would have been a good experience. But, you know, sometimes when you see this in front of your face, and you see a ship landing in front of you, you're in shock. You become so shocked that you can't know how to act or feel, you know. So your human feelings get in the way, you know. I didn't know at that time how to separate my feelings on that.

JP is here acknowledging that at the time this incident occurred in 2018, he still was not good at separating feelings such as shock in a contact situation from the apprehension he felt of being under surveillance by Men in Black/CIA personnel. The shock and apprehension combined to prevent him from making a choice that would have helped him learn more about the Nordics and their spacecraft. When the opportunity arose again with another invitation (see Chapter 3), JP was in a different emotional state, as he explains.

> JP - I had another incident when they invited me inside the ship to drink that blue liquid, and I went right away. I took that chance, and I did it, you know.

> MS - That was a later mission.

JP - And I actually, yeah, I took a picture of that ship when it was taken off, and you can see the reflection of the trees when the ship was heading up. So the bottom part of the ship was the same green as the trees and all that.

MS – Yeah, and we released those photos. You took several photos, and we released them on exopolitics.org back in the day. So people can still go and watch what's the little video I made of that on the photos.

The photos and video that JP is referring to are discussed in detail in Chapter 3, where some of the photos are included. The panel discussion continued as follows:

MS - So what do you know, Elena, of Nordics or Galactic Federation personnel that are working with, say, different branches of the military and have their own little squadrons?

ED - Yeah, so that will be, again, the Earth Alliance. That will be the programs that started in the early 1950s, or when William Tompkins testified of work with the Federation, you know, working on technology exchanges. So, at that moment, the Earth Alliance started to be created. That was the start of the Earth Alliance, the beginning when the Galactic Federation helped humans of Earth to build their own defense system, defense fleet, against the invaders. So that would be at that moment. And these programs have been going on and have developed. And there are personnel of the Galactic Federation of Worlds sent to Earth on some particular missions or jobs.

It can be many years in "jobs" in some organizations. They have been working with the US Navy, that's for sure, since the beginning [1940s]. I also know that other parts of the American military, such as the US Air Force, have joined the Earth Alliance. But what Thor Han told me once about the Air Force, the US Air Force, is that it's very compartmented. There are black programs as well in the US Air Force, but there are also cells. There is actually one cell that is working with the Alliance. So sometimes you're talking about the Air Force, but you don't know which camp you are talking about. So, what JP describes is absolutely totally the work of the Earth Alliance Corporation.

MS - You've described being on at least four missions to underground Ant People kingdoms where you've seen the Ant People. In your first mission, you described seeing a sarcophagus with a giant inside that was being protected and kind of worshipped by the Ant People and that he was a king, right, but he was in some kind of stasis, and you didn't know who the [Giant's] name was. Actually, part of your mission was to find out who that was.

Later, I was with Elena, and we got information from the Galactic Federation that this was Ningishzida, an … Anunnaki scientist who had stayed behind for a time when the Seeders would return or when Ea/Enki would return and that he would awaken and share the knowledge, well the ancient knowledge. And he was the kind of sacred protector of trees and plants. He was an alchemist, a master alchemist. Now you described going, in one of your missions, to an underground Ant kingdom where there was a king, and you met the king, and he talked to you, and they were very interested in the arks, and then in your last mission you described when you went there that the Ant king had died and the Ant People had left, and all that was left in that area were humans and the Ant people had gone to another realm with the king who had awakened or the giant who had awakened. So, do you want to elaborate on that?

JP's first three missions to the Ant People kingdom are described in *US Army Insider Missions 2*. The 4th mission was described in Chapter 8.

JP - When we started going to the other realm, It was a long walk through [an underground tunnel system] that you go through to that [location]. I was feeling kind of dizzy and not sick, but I was not feeling right about going. I thought I was not ready to do that particular mission. But they wanted me to do it. It's similar to the feeling I felt when I was just overwhelmed by what I was seeing. My emotions were getting involved in the situation. So, I did not really want to go to it because I didn't know what I was going to feel or what was going to happen to me.

See, my body is going through a lot of changes. You know, healthwise. So, personally, I didn't want to put more pressure on my body to go through these experiences and to go through these different realms, if you want to call it that, because when you go through these different realms, everything moves differently. Similar to when the Chinese were behind the Indians, the Mexican Indians, you can see that time moves differently ...

MS - Just for the people who don't know, you're referring to one of the earlier Atlantic Space Ark space missions where there was a group of Aztec Indians that were singing and chanting "A Kuria Matte" walking into the ark and the Chinese soldiers that were there tried to follow them, but they kind of like got stuck in molasses because there was a temporal effect and you can only get through that by raising your frequency like the Aztec Indians.

JP's first three missions to the Atlantic Space Ark were multinational and included scientists and military personnel from different countries.[96] The above incident happened during his first visit there which involved US and Chinese military teams, who were accompanied by two Aztec Indians and an interpreter. JP described how the Chinese got very angry when they experienced the stuck-in-molasses temporal effect that stopped them from following the Aztec Indians. The Chinese mistakenly believed the problem was being caused by the Americans, but JP explained that progress through the arks was a matter of frequency.

JP – So, I was feeling this temporal effect there in that different realm. And I did not want to go through that because of my health issues that I'm beginning to have. So, I decided not to do it. So, they gave me the drinks. They gave me everything to prepare my body for the trip. And I decided not to do it. I decided to go back, "Take me back". And they were OK with that. They were, "OK, so next time you come, you'll be ready, right?" It's like, "Yes, I'll be ready. I'll be ready physically and mentally ready to do these types of missions."

MS - Okay, so before we address the awakening of the giant and the role he had in replacing the Ant king who had died, I just wanted to get Elena's feedback on the toll on the human body when you're doing these kinds of missions because people think, well, JP, while you're doing incredible missions, I wish I could be part of that, and it would be so fantastic. You must be in top health, but it's actually the opposite. You want to explain.

ED - Yes, people notice I've lost weight, you know. It's very straining on the body. I'm not going into the underground, into the Inner Earth, but I'm going to space, as well, like you, JP, and it's very straining on the body. It's physical, and it's also emotional and nervous because you change environment…. I'm teleported. Teleportation is straining the body. You don't feel really well afterward. You need to recover because I'm not really used to it yet, even if I'm teleported regularly. Changing also frequency, you know.

When you go to Inner Earth, you go into a different frequency, and this also is very difficult on the body. You can feel dizzy. You can feel weird, and it's very tight afterward. You need to recover. When you go to space, you change into a [different] electromagnetic field. You know, these bodies we occupy are made on Earth. We have the minerals of the Earth, the fluids of the Earth, the water of the Earth, everything that comprises this body is fabricated on Earth. So, we are in tune with this planet. The heart beats in resonance with the Schumann resonance. When you get out of the magnetic field of the Earth, the heart panics because there's no more resonance matching, so you're on your own, and the body just is stressed, goes into a shock, a stress. What magnetic field should I tune in? So, of course, the magnetic field of the ship, if you are in an ET ship, but it's training. At one point, I developed heart problems that were fixed by them, but it's very strange, so not everyone can do it, you know? JP is a strong military man with military training. I'm not, so it takes a physical toll. Yes, it's very hard.

JP - And this training helped me out a lot, you know, in the military. Maybe that's why they decided to put me in the military to do these types of [military] training because I wasn't exercising every day in the beginning when I was doing these missions. I wasn't eating right. So, going through the military and basic training, I guess I train my body to eat right and work out every day. [This] is part of our daily life. Military life is working out every day.

I wasn't doing that [as a civilian], but I was doing that while being in the military. So, I was physically and mentally [strong]. My body was physically good enough to withstand the strains of the missions. So, I kind of understand where she's coming from, you know. It's quite tough on your body, on your eyes. Yeah, you can become blind if you don't wear the [protective eye] shields. Right. Your bone density goes down. You lose muscle. You lose spinal fluid. You start developing, like, arthritis. And there's a lot of people in the military that if you do the studies when they get out. They all of a sudden get arthritis. Their back starts going bad. You know, they start thinking in a crazy way, and they stop [remembering]. They get Alzheimer's, or they get Parkinson's disease because military people, they go through a lot of shit in their missions. And, you know, a lot of people, I'm happy that a lot more people are talking, started talking like that general, he started saying, yeah, they, we got, we got beings walking among us. You know, we got ...

MS - Admiral Tim Gallaudet. Yes.

JP - We got David Grusch, who recently came out, too, in the last year [2023]. So, like you said, Doc, this is coming out little by little. A lot of people know more about these types of missions or situations that the military or civilians are going through.

ED - I'd like to comment on the diet as well. I was asked to change my diet to be more able to withstand higher densities and higher electromagnetic fields so that I wouldn't get sick, and I would be more energetic and dynamic once I'm there and recovering better once I'm back. And also about the eyes, that's right, most of these ships are very bright inside. So, you have a shock, and it always takes time to adjust. When you finally adjust to the luminosity, it's hard because it's straining on the eyes. Yeah, I can confirm.

MS - So now I want to come back to your fourth mission or the last mission to the Ant Kingdom. You learn, you're told by the Ant People that ... the Ant king has died, that you had met on a previous mission. You're upset about that, that the giant has awakened and moved into a new realm, and that the Ant people followed the giant into the new realm. Do you want to explain what happened?

JP - Quantum physics, when there are two particles in two places at once, is similar to these different realms. When you visit these realms, you still leave a part of yourself back in the other realm. It's like when you look at yourself in the mirror; it's an analogy. When you look at yourself in the mirror, you see yourself, right? But you see yourself in a different space, a different realm. So, it feels kind of weird when you try to enter into this different realm. I did not go completely. Remember, I came back out because I didn't want to experience that because that takes a big toll on your body. And I'm sure they wanted me to go through it to bring back information for the military. I decided not to. Maybe later on, I will go through this mission, not being in the military, and do a nice update because the way I felt was always welcome to be there.

JP is here referring to his upcoming honorable discharge from the US Army, which occurred on August 31, 2024. He was told that after his discharge, he would still be sent on missions and that new opportunities would come up for him in terms of working for military contractors or other military services recruiting him. I will discuss this more in the final chapter.

MS - ... Explain how you were told by the Ant people that their king had died and that the giant had awakened and had taken the Ant people to a new realm.

JP - They were telling me, like, in a telepathic way that the Ant king died. It was a sad situation for me because I had personally talked to him. I remember talking to this king. And when they told me the news that he passed on, I felt like they were telling me that he was not with us in this realm. But he's still alive somewhere else. So maybe Elena can elaborate more with consciousness, of how that might work when you pass on and you go to another place in time.... They were telling me, the Ant people, that the other king was awakened and that he was in charge, that he was working already on doing different things in the other realm. And I was excited in the beginning when they told me that because ... I really wanted to go and see who this other guy was, who this king was, and he had awoken. And it was really amazing to hear that first sarcophagus that I saw, that king had awoken. I felt happy, I felt amazed, and I had a sensation that he was calling me to talk to me. You could feel it, you could feel that, I don't know, it's like a feeling that they communicate with you, but you can't really work out the words, but you know they're trying to communicate with you.

By the way they, you feel the frequency, their frequency. But yeah, I wish I would have gone in, Doc, but I did not go in and meet that king, the king who has awoken in that realm. But I knew, I felt his presence, I did feel his presence. You could feel that he was awake. It was the same feeling that I felt when I got close to the sarcophagus, you know, that same feeling, the same heart vibration that you could feel. So you knew that he was awake and alive, you know, and walking around in that particular realm.

MS - So what do you say to all of that, Elena?

ED - What I think JP describes is a shift in a higher frequency. The Inner Earth people live on several different layers of frequencies. As you go towards the center of the planet, the frequency becomes higher. So, when there's a danger, they can just go deeper into this higher frequency. And they can either go there or shift physically. We know there are frequency gates you can pass through. Or they can shift their consciousness to other bodies there. They would seem to die in the first place, transfer their consciousness, and awaken in another place. They can do that as well. Regarding the giant, his soul, his being, is an Anunnaki, an Anakh. So, he may as well have transferred into a higher density with the others. Or he may as well, because the world is not ready yet, have gone back to the Nibiru ship. We can say those are the possibilities.

What Danaan described here as the Inner Earth having different densities that become higher the deeper one goes underground matches the revelations of Radu Cinamar (a pseudonym) in his book, *Inside the Earth*.[97] In it, he describes the geophysics behind different density layers beneath the Earth and why scientific instruments have been unable to identify this. As Danaan explained, the deeper one goes into Earth's interior provides safety from threatening events happening on the surface that would not extend into higher densities.

MS - So what do you think? I know we had that incredible discussion and discovery when we were driving together in Tennessee, and we found out, holy cow, this is Ningishzida. You want to maybe explain that.

ED - Okay, let's remind [everyone about] this little episode. Yes. JP was telling [us about] his first encounter with this sleeping giant and the tree, this tree dripping liquid that was like a fountain of youth. And that was an amazing story. And we were driving the car, and I said, "Oh, I'm gonna ask Thor Han if he knows about this giant." And he said, "Yes, it's an Anakh, or Anunnaki, so that he was ... one of the seven scientists, of Enki scientists, which now we know under the name, the seven Apkallu, Sumerian name, the Apkallu."

In many Sumerian cuneiform texts, the "Apkallu" are described as seven demigods or sages or kings that provided wisdom and guidance to the rest of humanity.[98] They are responsible for the "tablets of destiny" and worked with Enki/Ea. After the Great Flood, they were banished by Marduk to the Abzu—the primordial void. According to information received by Danaan from Enki/Ea, what happened was that after the defeat of the Enki/Ea faction of the Anunnaki, seven of Enki's chief scientists decided to remain behind on Earth and go into hidden stasis chambers and wait for the next age to arrive. Danaan claims that the Return of the Seeders and Enki himself has led to the sleeping Anunnaki scientists, the Apkallu, reawakening.

> ED - Anyway, he [Thor Han] says, "Yes, his name was," and he says something I had so much difficulty transcribing. I told you, he says something like, "nin-gish-ta, nin-gish-za, nin-gish-za, something like this, nin-gish-zi-za, so nin-gish-za maybe," this was difficult. And we went back, I went back to the hotel, and I did some research on the internet, and you did as well on your site, and we just contacted each other. Oh, my goodness! We found out [it was] Ningishzida, one of the seven Apkallu, who is the Lord of the Good Tree. Ningishzida means the "Lord of the Good Tree." Which tree? The tree of life and of the underground waters. And he is also [known] as Quetzalcoatl and Djehuti or Thoth. That was a wow moment we will never forget.
>
> MS - Yes, that was really quite amazing. And JP, you also described that you did not do a mission but were given information that another one of these giants had been found. I believe it was in the city or

the remains of the ancient city of Nippur. We got information that this was actually another Anunnaki scientist, Aruna, who was the chief Anunnaki engineer of the space fleet. And so, you want to maybe just explain that information, how you got that information, that another sleeping giant had been found in this ancient area of Iraq called Nippur.

JP - The way I got that information is because when we were visiting the realm of the Ant People, I noticed that every giant has their own army. And they have their own soldiers protecting that realm, protecting that place. So when I heard about that sleeping giant over there and also that they have a sleeping army that awakens when it's disturbed. If somebody disturbs the realm where the sleeping giant is, the army will awaken.

What army? I don't know what type of species or underground civilization this is. But as we know, there are different types of, well, Inner Earth civilizations. We've got the Ant People. We've got human beings that look like us all around the world. Different types of Inner Earth People that look different and [some] that look like us, ... they look [like] a mixture of Indian and Oriental. They look more pale than us, and there's different types of Inner Earth civilizations protecting these different realms and all that. And also over there in Ukraine, I think that's why, I guess, the Russians, they have [occupied] the area where the sand dunes [are], you know. What do you call that place?

MS - Oleshky Sands?

JP - Oleshsky Sands, they're still in charge of that. So. That's how I got the information because of the armies and because of the civilization that protects the realm. So, over there [Nippur, Iraq], they also have a sleeping army, If you read the story of that particular sleeping giant over there.

ED - Makes sense.

MS - Now, I remember talking to you about that, and you did a communication, I think it was, I'm not sure if it was with Thor Han or Prince Ea, and you said that Aruna because he was the chief engineer for the Anunnaki space fleet, he had knowledge about advanced space technologies that were way in excess of the secret space programs, of the Solar Warden, and even of the Nordics.[99] So, do you want to recapitulate on exactly what it was that you were told about Arunna and his vast knowledge of space technologies?

ED - Yes, Aruna was the chief engineer of Enlil's fleet, and he was in charge of getting ships produced. You know, the Nibiru is a mothership, it's huge, it's nearly as big as the moon, and it has fleets inside, [including] a lot of scout ships, or fleets, for war combat. And Aruna was in charge of producing these fleets, or repairing them, or getting them, you know, fit, and he would also manage the, you can say, the astroport, the port for the ships, in Nippur, in the area of Nippur, because Nippur was the headquarters of Enlil. And Aruna, so Aruna was his chief of fleets, I would say, I don't know how to translate this.

MS - Chief Engineer.

ED - Chief Engineer, and what is quite impressive is that when I got this information we always do research. When you got the information, and I think it's you who found out that there was in India an Aruna who was the chief charioteer of a god, then Aruna, the chief charioteer of the gods

In the Vedic religion, Aruna is the charioter of Surya (the sun god) who was the father of Karna, the chief rival of Arjuna in the Mahabharata war. Aruna is the brother of the deity Garuda, who occupies a prominent role in Hindu pantheon.[100]

MS - Correct, yes. So, there's an allusion there to this ancient mythical figure who was exactly playing this role.

ED - And what is very important is that Aruna switched sides. He went on Enki's side at one point, and he joined his followers.

MS - Well, that's a kind of like a testament to, I guess, the character of the Anunnaki as a civilization, that there was this ability for leading figures because I think Ningishzida, as I recall, was he a son of Enlil or a closely associated blood-kin of Enlil?

ED - He was his mentor. He was older than Ea. Ningishzida was older than Ea.

MS - I'm sorry, I got confused, Aruna, yes, Aruna, wasn't he a son of Enlil? Or somehow, they were related, and they could shift allegiances, so there was this fluidity in Anunnaki society, which was very interesting.

ED – Yes ...

MS - Okay, so this is the last topic because we do want to keep it short. So, your most recent mission happened on July 4 of this year, 2024, and you described going to a medical facility on a military base, and you met three other soldiers. And at first, they didn't know you, and they thought, well, who is this guy? Who's this stranger? And they were very dismissive of you, saying, how come this guy is with us, and he shouldn't be here? We're trying to do these kinds of missions. Apparently, it was something to do with an upcoming mission to Saturn, and then all four of you went into this medical facility, and there was a table, and you had to place your palm on it. So why don't you explain what happened?

JP - So yeah, we got this update. This [update] is going to be number 35, where we put the hand on the tables, right? Cause I know they're going to hear the mission. They're going to hear the story.

MS - You can say it again, just for people who may need reminding.

JP - They took us into this facility, and they lined us up. One of the guys was like, "Who the heck are you? You didn't go with us on all the training. You didn't do any of this. So why you're here?" So, I guess they did have a fourth guy, but he did not make the cut, I guess, and they added me into their group. So they were sad that they lost that particular partner. And I took his place.... They put us into this room, which has a metal table, and I put my hand on the metal table, and you could see numbers that come up. It tells you your oxygen level, the type of blood that you have, your DNA, and all that, and then on top of that, there are other numbers, medical numbers that I did not know, and on top of that, there was a color code. It was reading greenish-yellow, fluorescent yellow. The other guys were reading like a bluish-purple.

So, the guy that was cursing me out saying that he didn't want me [there], he looked at my color, he's like, "Oh crap, look at his color, man. We never see that. That's freaking rare." So, the color they explained to me that the colors meant how far you have been [in outer space]. The green meant that I've been outside of the solar system. That's how far I've been. The bluish-purple meant, like, as close as Jupiter and all that, but the green means way farther, to like a farther place. So, when they saw that, they started respecting me a little bit more, treating me better, and all that.

And they're like, "Whoa, we can't believe we saw that color. That's really rare. And sorry, man, dude, I'm sorry." So, they treated me in a different way, and they told me that. When I came back, the guy who gives me the green light, he says that I had probably been on this mission before. The reason is that color. So it was an amazing experience.

MS - Yeah. So, Elena, do you want to respond to that?

ED - Yes, well, when you leave the electromagnetic field of the Earth, you are going to adjust automatically to any new magnetic field that you encounter on your trajectory. When you are in an ET spaceship, this doesn't happen, you just tune into the magnetic field of the ship, and that's the trace you will find in your own personal magnetic field that can be retraced. There are traces of a different frequency, but that will be the frequency of the spaceship. But when you travel in human-made, Earth-made spaceships, you don't have the same electromagnetic field protection. So, you are going to catch the electromagnetic field of the planets you pass by or the places you go to in space. It is all recorded in your own personal magnetic field, so you can find these magnetic signatures. I'm convinced that's what JP's analysis of the colors was showing.

JP - Yeah, you can also explain [about] Jean Charles, what he told us. You can explain what he found also.

MS - We had a conversation with Jean Charles Moyen, and he described having a similar experience where he placed his hand on some kind of metallic device and said that it would give a readout. And depending on the color. He said it was two colors, green and blue. And depending on the readout, that would tell a lot about your ability to be able to go out into space on missions. Yeah, it's amazing.

I discussed Jean Charles Moyen's response to JP's account of the hand-reading machine that could evaluate one's suitability for space travel in the previous chapter, which contains the full version of JP's account of the upcoming/completed Saturn mission.

MS - That mission was to Saturn, to the rings of Saturn. You described that you were being prepared for a mission to Saturn and that you don't remember that mission.

JP - No, they're telling us in a briefing that we're going to meet up with some other ET race, and the ship is going to rendezvous and communicate with each other. But by the way, my handler, the guy that gives me the green light, he told me that who knows, you probably already went on this mission, and you're just getting briefed on what happened. Or what you did already, but in the briefing, they made it seem like we're going to go through the mission. But I don't know if I have been on the mission already. The reason why I was showing green in the table [may be because I already have been on the mission].

MS - Is this some kind of time paradox thing, or are we talking about the anti-telephone effect where you go on missions because you've already been and you've returned? This is something that Tony Rodrigues described in his presentation [at Valance, France], the anti-telephone effect. The Dark Fleet would do that that they would send out missions... [but] they would first wait, they would use this anti-telephone effect. Before launching the mission, they would wait for themselves to return from that mission. Once they see themselves returning, then it's like, "Okay, we'll go on that mission because it's safe." And if they didn't see themselves returning, then they'd scrub the mission because, obviously, there was some danger there that they would encounter. So, I don't know, is this maybe an anti-telephone effect?

Tony Rodrigues described how the Dark Fleet would perform this temporal maneuver (aka anti-telephone effect) before starting missions in his 2022 book, *Ceres Colony Cavalier*. Rodrigues claims he spent 13 years as a slave in a merchant ship that performed deep space missions for the German-led Dark Fleet.

> JP - It could, it could, it could be an anti-telephone effect. It could be, but I don't have any recollection of it. [I have] no memory. I know I have told you about 35 missions, but in between these four years (2020-2024), I don't know how many missions I probably did that maybe affected me in a certain way. That's why I'm not allowed to remember these certain missions.
>
> Remember, I'm still part of the military, so I'm still serving. [JP's military status changed on August 31, 2024, nearly two months after this interview was recorded]; I love the United States of America for giving me the opportunity to serve. And I'm really proud of our military, and what we're going to [do], you know. There are secrets that we still have to keep, and we still have to keep them for the security of the United States of America. And I'm not scared to say that because I am part of the greatest nation in the world, I believe. Of course, my wife, she's from Brazil, so I love Brazil too, and right now I'm in France, so France is a beautiful, beautiful country, Elena. Europe is beautiful, but I do serve the United States of America, one of the best countries in the world and more advanced in technology, I believe.
>
> MS – Yeah, so there's a mission to Saturn, the rings of Saturn, or the Saturn that JP has done at some point that he can't recall now, but he was being tested for that mission, and then he finds out from

his handler, who's a senior military officer, that that mission may have already been completed. So [Elena], what do you say to that? I mean, about Saturn and Earth Alliance missions there?

ED - Oh my, well, Michael, people who follow my Star Nation news know what I'm going to talk about. And at the time, if I can't recall either, I told you JP is going to be called on a mission to Saturn. Do you remember?

MS - Yes, I do.

ED - So you will hear, he will tell you, that is, I didn't tell you [JP] because I don't want to influence anything.

MS - Yes. Exactly.

ED - I told you. And yes, I may have mentioned it in one of my Star Nations News because...

JP - But you never said JP [on Star Nations News]. You only told that to Doc.

ED - I may have mentioned JP; we may hear JP telling us about a new mission to Saturn. Let's see, you know, I may have said it or not, but I don't remember.

JP - ... I only stay on my social media, but I try not to go on different other [channels so as] to not get influenced by information, you know.

ED - I think I told you about it.

MS - You did, you did tell me.

Elena Danaan sent me a private message on WhatsApp on January 17, 2024, where she related information from Thor Han Eredyon about Titan and Saturn being handed over to the Earth Alliance as a 'gift' from Prince Ea (aka Enki). She speculated that "I wouldn't be surprised if JP gets to go on a mission on Titan, or Saturn, in the coming times" (see below figure). What made this message very significant was my skeptical response that JP was not doing off-world missions with Space Command at the time. Based on JP's experience and what he was told by his handler, he had indeed, done a mission to Saturn not long after this formal transfer of Saturn and its biggest moon, Titan, to the Earth Alliance in January 2024.

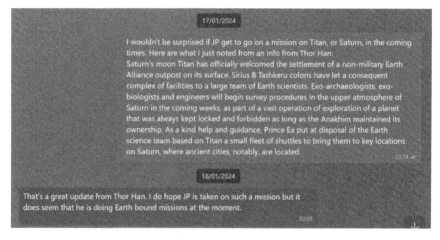

Figure 27. Elena Dannan communication on JP and a future Saturn mission

ED - But what I said in my Star Nation News is that a few months ago, I got the information that the Earth Alliance invited a team of scientists. There were topographers, biologists, and other types of [scientists] that were brought by the Earth Alliance and military to protect them and to bring them there to Titan, a moon of Saturn. They were received there, and they were given quarters and facilities to be brought to Saturn to study ancient floating cities. Because on Saturn there was a group, a very ancient group called the builders, who left floating cities and anti-gravity cities.

But that was hidden from everyone because you could never access Saturn because it was previously in the custody of Enlil and Marduk, the bad Anunakis. You couldn't approach. So recently, these bad Anunnakis have been expelled from our star system, and Enki or Ea came back in September 2021 and gave back the keys. He said, okay, on behalf of the Anakh Empire, I am giving back the custody of Saturn to the Earth Alliance because Earth is the more advanced indigenous civilization in the system. Then, Saturn was open to visits by the Earth Alliance, and they saw these cities. So the Anakh, or the Anunnaki, let's call them that for better understanding…. Enki's team said, well, we're going to help you and give you facilities on Titan and guide you to these cities. So there's a detachment of an Anunnaki group from the Nibiru, Enki's personnel, who is helping and guiding the Earth Alliance scientists and the militaries that are protecting them to these different cities.

So, they go on these shuttles with these Anunnaki people and personnel, and they are brought to these cities. What is interesting is that I've seen these cities from afar, and I had a visit as well. They are like ruins, but in the upper atmosphere of Saturn, like big ruins of floating cities. And I asked this question, "If they're that ancient, how does the mechanism, anti-gravity mechanism still work?" And I was told, "There is no anti-gravity technology. It's the materials by themselves that have no mass."

So they float; it's totally safe. So, the Earth Alliance scientists are studying these cities.

JP - So the gravitational pull of Saturn will not affect [them].

ED - No, it will not affect [them].

JP - So the structures that are around the rings of Saturn. I know there's a book out by some guy, he found structures.

MS - It's called ... the *Ring Makers of Saturn*.

Dr. Norman Bergrun, a distinguished scientist who worked at the Ames Laboratory (NACA) and Lockheed Martin for 24 years, wrote the *Ringmakers of Saturn* in 1986. In it, he used 45 photos from NASA's 1980 Voyager 1 and 1981 Voyager 2 flybys of Saturn that showed very large spacecraft, including a 7000-mile-long cigar-shaped craft, orbiting the rings of Saturn. In the Preface to his book, Bergrun wrote:

> Presented herein are pictures of immensely large, enormously powerful extraterrestrial space vehicles located in the vicinity of Saturn and its moons. These photographic revelations are reinforced by and are consistent with scientific data extending over centuries as far back as Galileo. The pictures have been obtained by the author using simple repeatable enhancement techniques applied to publicly available NASA photographs from Voyager one and two flybys of Saturn.... Identification of extraterrestrial vehicles apparently possessing ancient historical presence in the solar system is a new discovery having many ramifications....

Magnitude of the Saturnian breakthrough would appear to be substantial. Saturnian space vehicles, strangely unusual in their great size and appearance, introduce a new and unpredictable variable into affairs worldwide. Compelling reasons exist for obtaining a much more complete understanding of these vehicles and the influential superlative intelligence behind them.[101]

Dr. Bergrun concluded that these giant spacecraft were responsible for making or using Saturn's rings in unknown ways. According to Danaan's information, these giant spacecraft photographed in 1980 and 1981 were being studied or used by the Enlil faction of the Anunnaki.

JP - The *Ringmakers of Saturn*, do you think they're involved?

ED - I haven't read that book, so I cannot pronounce an opinion. I haven't read it, but there are a lot of structures. In the rings of Saturn, you could not access the inner ring from the inner ring towards the surface of Saturn. It was forbidden because it was Anunnaki territory, but the rings were accessible to anyone who would come and maybe put a mining station there or an observation force, but honestly, the Galactic Federation, they didn't want to get that close to the Anunnaki territory because they were troubled. It could be dangerous, yeah. They were troubled.

JP - They told us it was a dangerous mission going there.

ED - That's it. Yeah.

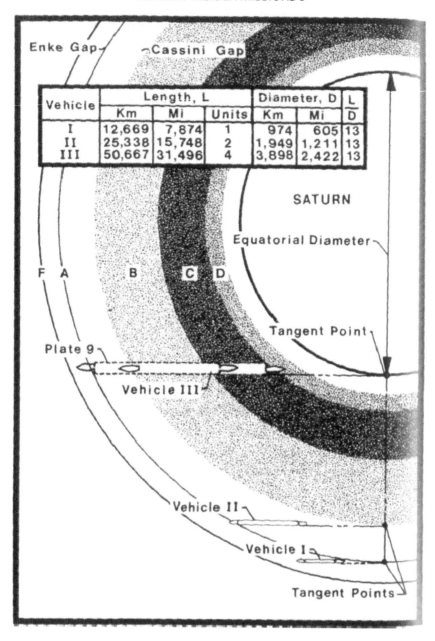

Figure 28. Spacecraft identified around the A, B, C & D Saturnian rings. Ringmakers of Saturn, p. 50.

JP - I guess it's because of the activation of certain ships over there that are involved with doing [things with] the rings of Saturn. So, I think we're going over there to rendezvous with these particular technologies that we were not sure how ancient they are. And it's quite interesting.

ED - Yeah, it could be the same thing. Yeah.

JP - So, can that be involved with the hexagon shape on top of Saturn? ...

ED - I do not think so because I think that it is a question of winds and rotation that creates this geometrical shape. I think it's a natural phenomenon from what I understand.

MS - Well, that really was amazing information. I want to thank you both. I mean, what you shared was incredible, and it was just kind of like unraveling the deep implications of the missions JP is on... This has been a wonderful conversation.

ED - Thank you, Michael.

JP - Thank you. Thank you.

The roundtable discussion was very helpful in unraveling some of the deep implications of JP's missions, and what lies ahead for him both in conducting in future missions and remembering more of his past missions. The question of future missions is particularly relevant given that on August 31, 2024, JP's military status changed significantly.

Chapter 12

Honorable Discharge from US Army and Future Covert Missions

August 30, 2024, was JP's last day in the US Army. He has been given an Honorable Discharge from active-duty military service and granted a life-long permanent disability pension for the injuries he has suffered during his service. JP's honorable discharge for medical reasons comes as no surprise. On many covert missions, JP has described the growing health toll on his body from his experiences, which involved exposure to different electromagnetic fields and radiation levels encountered during space travel, visits to different densities and dimensional realms in the Earth's interior and underwater, and interactions with advanced technologies from both ancient civilizations and visiting extraterrestrials. As explained in Chapter 11, the injuries from exposure to such radically different environments and technologies are cumulative and can be difficult to overcome without significant medical assistance.

While JP has been given access to different healing tools by the Ant People, Nordics, and the US Army, these have not proved sufficient for his full recovery. Another key factor that has impacted JP's physical health is all the COVID-19 vaccine shots he was required to take to remain in the US Army at a time when this was mandatory. Soon after receiving the COVID-19 vaccine, he began experiencing medical problems with his heart that significantly impacted his regular military duties. None of the Army doctors have been able to help him, and he has been denied access to advanced MedBed technologies used in the world of covert operations and secret space programs.

It may appear puzzling as to why JP has been denied access to advanced healing technologies since he has performed covert off-planet missions along with missions to the Inner Earth on behalf of the US military. This denial of access to advanced healing technology can be traced back to the tug-of-war battle between different factions within the US military and the Intelligence Community, specifically concerning JP's disclosures. While "White Hats" encouraged and facilitated JP's release of what he witnessed during different covert missions, "Black Hats" opposed JP's disclosures at every opportunity. He has also been exposed to intimidation, harassment, and even assassination attempts.

In *US Army Insider Missions 1*, I explained the secret technology agreements reached between the US Air Force and a faction of Nordic extraterrestrials who had first made open contact with JP in 2008. Conditions of this USAF–Nordic agreement included that JP would be protected from "Black Hats" and would be allowed to reveal details of the different classified programs he was exposed to by USAF operatives. For example, in 2017, JP was encouraged by covert USAF operatives to take photos of classified antigravity craft operating out of MacDill Air Force Base in Tampa, Florida; home of Special Operations Command. Meanwhile, JP was harassed by Men in Black and CIA operatives who tried to block the release of photos and videos he was capturing of the antigravity craft operating out of that base.

JP had been encouraged by USAF operatives to enlist in the US military as this would allow him to be given access to highly classified programs, which his civilian status denied him. In *US Army Insider Missions 1*, I described two instances where JP was kicked out of a classified program by senior security officers despite covert USAF operatives arranging for him to be given access. JP had been told to enlist in any military branch because he could easily be transferred as needed for covert assignments on a temporary basis. Consequently, JP enlisted in the US Army in late 2019, and began boot camp on January 7, 2020. After graduation, JP was trained to be a Quartermaster and Chemical Equipment Repairer, and he also received paratrooper training, which further equipped him for his coming special missions. Later, JP would be called away when needed to perform covert missions out of a major military base in Florida. He described the process as one where he would receive a coded message on his mobile phone and would then follow the instructions given. At the time of writing, JP has performed over 35 missions, although he is now aware that this number is, in fact, far greater. It has been unsettling for JP to learn that he does not recall an unknown number of missions due to the memory-wiping process he has undergone, as explained in Chapter 10.

The senior USAF officer who greenlighted JP's public mission reports has informed him that while his regular US Army career is going to end with his honorable discharge, he will still be called upon to do covert missions when needed. JP was further told that he would likely receive offers from different corporate contractors, government agencies, and other military services to work with them on classified technology programs. This process has already begun to happen, which is why JP has decided to maintain his anonymity until he is clear about his future career path. He also has decided not to reveal details of the Army unit he served with and the military base he was stationed at for similar reasons.

While JP plans to maintain anonymity, using only the initials of his first name (J) and surname (P), he has given me permission to release redacted versions of some of his official documents, which give more details about his military service than those previously released in *US Army Insider Mission 1*. The first document presented here is the official retirement letter from the US Army issued on August 30, 2024, that refers to him "having served faithfully and honorably."

Figure 29. Certificate of Retirement

JP has earned achievement medals from the US Army and US Navy/Marines. The US Navy medal is especially significant. This medal refers to his serving as a translator while demonstrating technical acumen in the training of military personnel from other countries. However, the Navy has also had a great interest in JP's covert missions to space arks and its own need to train personnel for similar future missions.

Figure 30. Navy and Marine Corps Achievement Medal

Finally, I include a redacted version of JP's DD 214, which is the official summary record of military service and discharge. The DD-214 confirms that JP served honorably from January 7, 2020, to August 30, 2024 – a total of 4 years, 7 months, and 24 days. He achieved the rank of Specialist (E-4) and officially served as a Quartermaster and Chemical Equipment Repairer (MOS 91J10). JP received three medals during his Army service. These are the Army Achievement Medal, the US Navy Achievement Medal, and the Army Good Conduct Medal. He also received the Army Service Ribbon and two badges.

JP's honorable discharge from the US Army opens up exciting new pathways for JP's future disclosures, and he remains staunchly dedicated to revealing the full extent and details of his covert military missions and contacts with extraterrestrials. The covert military officers who have been working with him have already told him they want him to do more missions and continue his disclosures with his new status as a US Army Retiree.

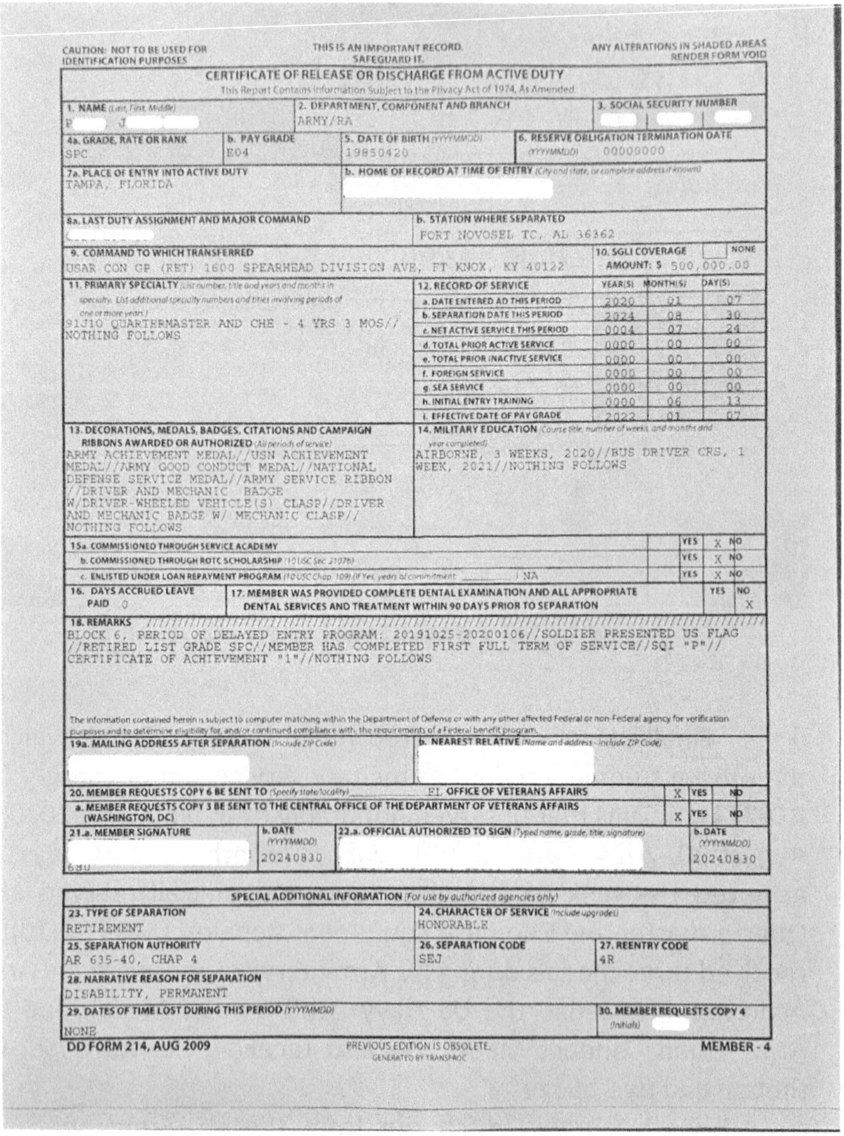

Figure 31. DD Form 214

HONORABLE DISCHARGE FROM US ARMY AND FUTURE COVERT MISSIONS

JP's impending new status has led to regular army personnel expressing considerable jealousy upon learning he would receive a life-long disability pension at nearly the full rate after serving only 4 years and 7 months without any obvious physical injury, such as loss of limb. In Luis Elizondo's new book, *Imminent,* he points out a new Pentagon policy, which has recently been implemented. It states that retirement status and benefits will be given to military personnel who have suffered medical issues from interacting with UFOs, at a rate of up to a 100% disability despite their length of service.

> Whatever technology makes UAP fly clearly generates a form of radiation that can be deleterious to living human tissue, but how do we account for the other weird things such as time distortion, perceived psychic ability, and so on? Is the radiation doing it, or is there something else at play? People would be surprised to learn that the US government has awarded multiple servicemen 100% disability in writing due to medical issues resulting from their close encounters with UAP. [102]

This is the precise situation in which JP finds himself, as I've described in earlier chapters. JP's unusually high percentage retirement pension is, therefore, a parting gift arranged by his covert military leaders, who are looking after his long-term financial interests and they will no doubt set him up for future related missions involving advanced flying technologies (aka UAPs) and underground civilizations. To those covert military officers and "White Hats" who made it possible for JP to receive this parting gift and who also protected him over the years in disclosing what he had learned, both prior to and during his military service, I offer my sincere thanks and dedicate this book to you.

With JP's retirement from the regular US Army, he can now dedicate himself full-time to different disclosure initiatives. JP may also decide to accept a position with another military service or organization to increase his knowledge and exposure to classified space-related programs. In this regard, he has already been approached with an offer to enlist in the US Navy, which he has been encouraged to accept by Alex Collier, who has his own military network of contacts. JP was advised that the Navy's Solar Warden program looks after its operatives better than the US Army. Also, the Navy offers better career prospects for covert operatives who often mix their missions with regular duties. Critically, JP would have access to advanced healing technologies, such as the MedBeds used in Solar Warden.

JP has also been advised that if he joins the Navy and is recruited into the Solar Warden program, he will work directly with the Galactic Federation of Worlds (GFW), which has formal agreements with the Navy, according to Elena Danaan. She says there are no formal agreements between the GFW and the USAF. The USAF instead has agreements with several Nordic groups, including the one that initially contacted JP in 2008. Joining the US Navy would give JP access to Galactic Federation personnel such as Thor Han Eredyon, who are working with the US Navy.

In addition, there is exciting new information that a more highly evolved faction of the Solar Warden program has returned from an extended tour of duty in interstellar space. According to George Kavassilas, the senior officers in this part of the Solar Warden program became spiritually enlightened due to their interaction with highly evolved galactic entities in interstellar space.[103] The admirals running this portion of the Solar Warden program are similar to Jedi Knights, and it is claimed they have just returned to our solar system and will play a critical role in the disclosure process that is about to be unveiled to global humanity. If JP does choose to accept a position in the Navy, then his experiences and skills will be a good fit with the Solar Warden program, especially this newly returned enlightened group of Jedi-like admirals.

At this point, it's not clear when and from which organization JP's future missions will be conducted. His current choices include working as an independent military retiree, working with an aerospace corporation with lucrative financial benefits, or joining another military service such as the US Navy. One thing JP has made clear to me is that he is dedicated to sharing whatever he learns in future missions and also from recovered memories of past missions. I look forward to reporting on these updates in future volumes of US Army Insider Missions.

US ARMY INSIDER MISSIONS 3

Acknowledgments

I am deeply grateful to JP, who has confided in me his incredible extraterrestrial and secret space program experiences for 16 years and given me photos and videos of the different types of antigravity ships and orbs he has encountered. During his US Army career, he shared direct experiences in covert space operations and missions to underground bases and submerged space arks, despite reprimands and financial risk. Up to his retirement from the US Army on August 30, 2024. JP shared much sensitive information about how the missions were conducted from a major military base where he worked, which he was able to verify through unredacted versions of documents he supplied to me. JP's actions speak volumes about his courage and commitment to full disclosure of UFOs, ancient civilizations, and extraterrestrial life.

I'm deeply grateful to an unnamed senior US military officer and other unknown military personnel who have greenlighted, assisted, or protected JP in sharing his firsthand experiences in missions both off, on, and inside our planet. In doing so, these anonymous military personnel have allowed the general public to get a rare glimpse into a highly secretive breakaway civilization associated with deep underground military bases, extraterrestrial visitors, cryptoterrestrial life, and incredibly advanced aerospace and healing technologies. Profound thanks to all of you!

I greatly appreciate Adam S. Doyle for creating yet another incredible cover that incorporates scenes from JP's missions. He has now created the covers for all three volumes that make up the *US Army Insider Missions* book series. Thank you Adam.

Thanks also to Dine D'ávilla for her graphic designs illustrating the two recipes JP was given by the Ant People for health and longevity drinks.

Finally, my heartfelt gratitude goes to my best friend and soulmate, Angelika Whitecliff, whose enthusiastic support and sage advice were instrumental in my writing, editing, and design of this book.

Michael E. Salla, Ph.D.
September 8, 2024

About the Author

Dr. Michael Salla is an internationally recognized scholar in global politics, conflict resolution and U.S. foreign policy. He has taught at universities in the U.S. and Australia, including American University in Washington, DC. Today, he is most popularly known as a pioneer in the development of the field of 'exopolitics'; the study of the main actors, institutions and political processes associated with extraterrestrial life.

Dr Salla has been a guest speaker on hundreds of podcasts, radio and TV shows including Ancient Aliens, Coast to Coast AM, Redacted, and featured at national and international conferences. His Amazon bestselling *US Army Insider* (3 volumes) and *Secret Space Program* (7 volumes) book series have made him a leading voice in the Disclosure Movement. He is also the host of the popular podcast *Exopolitics Today* with over 175,000 YouTube subscribers. His main website is Exopolitics.org .

Also by DR. MICHAEL SALLA

US Army Insider Missions 2: Underground Cities, Giants & Spaceports
– Book Two of the US Army Insider Missions Series –

US Army Insider Missions 1: Space Arks, Underground Cities & ET Contact
– Book One of the US Army Insider Missions Series –

Galactic Federations, Councils & Secret Space Programs
– Book Seven of the Secret Space Programs Series –

Space Force: Our Star Trek Future
– Book Six of the Secret Space Programs Series –

Rise of the Red Dragon
Origins & Threat of China's Secret Space Program
– Book Five of the Secret Space Programs Series –

US Air Force Secret Space Program
Shifting Extraterrestrial Alliances and Space Force
– Book Four of the Secret Space Programs Series –

Antarctica's Hidden History
Corporate Foundations of Secret Space Programs
– Book Three of the Secret Space Programs Series –

The U.S. Navy's Secret Space Program & Nordic Extraterrestrial Alliance
– Book Two of the Secret Space Programs Series –

Insiders Reveal Secret Space Programs & Extraterrestrial Alliances
– Book One of the Secret Space Programs Series –

US ARMY INSIDER MISSIONS 3

Kennedy's Last Stand
Eisenhower, UFOs, MJ-12 & JFK's Assassination

Galactic Diplomacy:
Getting to Yes with ET

Exposing U.S. Government Policies on ET Life

Exopolitics:
Political Implications of Extraterrestrial Life

Index

Adair, David, 86, 287, 288
Ahel, Pleiadians, 297
Alabama, x, 2, 205, 233, 234, 235, 239, 243
All-domain Anomaly Resolution Office, 27, 242
Amazon, 339
Ant King, viii, 203, 207, 215, 230, 277
Ant People, viii, x, 2, 3, 48, 53, 93, 94, 203, 207, 208, 209, 210, 212, 216, 217, 219, 222, 226, 231, 246, 277, 301, 302, 306, 311, 327
Ant person, 211, 214, 216, 217, 221, 224
Antarctica, 182, 235, 341, 346
Antarctica's Hidden History, 341
antigravity craft, 21, 51, 246
Anunnaki, 3, 74, 231, 281, 301, 309, 310, 311, 312, 313, 321, 323, 325
Area 51, 86, 244, 245, 287
Artemis Accords, 3, 248
Aruna, 312, 313
Arunna, 313
Asteroid Belt, 29
Atla, 284, 285, 290
Atlantic Ark, vii, 8, 25, 27, 28, 57, 58, 75, 83, 95, 99, 111, 160, 281, 284
Atlantic Ocean, 1, 23, 35, 61, 87, 98, 100, 139
Atlantic Space Ark, vii, 1, 5, 6, 11, 13, 14, 15, 16, 19, 29, 31, 32, 35, 61, 74, 75, 79, 81, 88, 91, 95, 111, 132, 133, 135, 138, 139, 164, 169, 170, 281, 282, 284, 288, 293, 303, 346
Aztecs, 14, 17, 91, 217
Bahamas, 19, 346
Bergrun, Dr Norman, 322, 323
Bermuda Triangle, 8, 19, 27, 35, 79, 81, 88, 100, 107, 132, 133, 291

Black Hawk Helicopter, 6, 24, 149, 151, 153, 179, 181, 182, 184, 199, 229, 255
Brazil, 59, 64, 72, 78, 109, 130, 132, 154, 175, 178, 200, 243, 248, 249, 290, 318
Britain, 98, 134
Carlson, Tucker, 250
Ceres, 29, 58, 143, 318
Chemical Equipment Repairer, 329, 331
China, 15, 58, 98, 109, 132, 134, 135, 138, 248, 341, 346
Chua, Stephen, 244, 245
Ciakahrr, Reptilians, 74
Collier, Alex, 79
Convention on the Protection of the Underwater Cultural Heritage, 136, 137
COVID-19, 327
crystal jewel, space ark, 1, 11, 13, 29, 35, 75, 93, 282, 286, 288
Dakota, North or South, 177
Danaan, Elena, 3, 13, 29, 73, 74, 86, 130, 156, 157, 217, 244, 245, 248, 279, 281, 287, 309, 310, 320, 323, 346
Dark Fleet, 317, 318
Deep State, 2, 33, 201
disability pension, 327, 333
donut-shaped ship, 6, 9, 16, 18, 22, 23, 24, 81, 83, 87, 89, 96, 100, 101, 102
Dragon's Triangle, ix, 27, 28, 109
Earth Alliance, 2, 3, 62, 111, 141, 177, 202, 293, 296, 297, 300, 301, 319, 320, 321, 322
Eisenhower, President Dwight, 342
Enki/Ea, 231, 284, 301 309, 310, 313, 320, 321
Enlil, 312, 313, 321, 323

Europe, 13, 18, 21, 78, 175, 178, 222, 318
Exopolitics, iv, 6, 37, 62, 82, 109, 112, 178, 204, 278, 282, 339, 342, 346
Exopolitics Institute, iv
Exopolitics Today, 339
Florida, 1, 2, 3, 93, 127, 146, 183, 203, 205, 233, 240, 244
flying saucer, ix, 37, 41, 42, 43, 51, 61, 63, 65, 77, 83, 157, 236, 237, 241, 242, 244, 298
France, 98, 248, 266, 279, 281, 294, 317, 318
Galactic Federation of Worlds, 231, 247, 281, 285, 296, 300
Galactic Spiritual Informers Connection, GSIC, 156, 173, 281
Gallaudet, Admiral Tim, 277, 306
Ganymede, 29, 261, 267, 274, 346
gargoyle, 185
Germany, x, 168, 248
Gray extraterrestrials, 191, 194, 195, 201
Grusch, David, 56, 148, 257, 306
Gulf Stream, 9, 10
Halls of Records, 74, 346
holographic technologies, 119
Honorable Discharge, 307, 327, 329, 331
Huntsville, Alabama, 235
Hurricane Lee, ix, 8, 9, 16, 19, 35, 57, 88
Hydronium, 253
India, 53, 64, 65, 138, 248, 292, 313
Inner Earth, v, vii, 1, 3, 21, 37, 48, 50, 51, 52, 60, 81, 93, 107, 116, 139, 160, 169, 188, 193, 196, 201, 222, 228, 231, 246, 258, 292, 304, 308, 309, 311, 328, 346
Intergalactic Confederation, 29, 74, 86, 284, 286, 287, 293
interstellar craft, 246
Italy, 166, 168
Japan, 27, 109, 157, 162, 170, 248
Jafis, 1, 61, 68, 79, 92, 105, 111, 128, 129, 244
Jupiter, 29, 58, 143, 243, 248, 261, 267, 272, 315
Kavassilas, George, 334

Kennedy, John F., 342
Kennedy's Last Stand, 342
Kirkpatrick, Sean, 27, 346
Loeb, Avi, 27, 346
MacDill Air Force Base, 3, 21, 146, 328
magnetic field, 228, 304, 315
Maldek, 74
Marduk, 310, 321
Mars, ix, 29, 74, 143, 213, 267
MedBed, 328
Men in Black, 296, 299, 328
Menger, Howard, 166, 167, 297
Mexicans, 14
Mexico, 72, 78
MJ-12, 342
MK-Ultra, 103
Moon, ix, 95, 143, 153, 154, 160, 164, 169, 246
Moyen, Jean Charles, ix, 19, 20, 266, 316
Navy's Secret Space Program, 341
Ningishzida, 203, 217, 231, 301, 309, 310, 313
Nippur, Iraq, 311, 312
Nolan, Dr Gary, 250
Non-Human Intelligence, 257, 277
NORAD, 23
Nordic spacecraft, 61, 233, 243
Nordic extraterrestrials, vii, viii, ix, 1, 2, 3, 5, 11, 14, 15, 17, 22, 27, 30, 35, 47, 48, 50, 51, 52, 53, 58, 61, 67, 70, 72, 73, 74, 75, 77, 79, 81, 87, 88, 90, 91, 92, 93, 94, 95, 97, 98, 99, 102, 105, 108, 110, 111, 123, 128, 131, 132, 139, 140, 141, 145, 147, 151, 152, 153, 154, 155, 159, 160, 161, 162, 163, 164, 166, 169, 172, 189, 190, 191, 193, 194, 195, 196, 201, 227, 231, 233, 237, 240, 241, 243, 245, 246, 247, 267, 270, 271, 288, 289, 290, 291, 292, 293, 294, 298, 299, 300, 312, 327, 341
Oleshky Sands, space ark, 311
Operational Camouflage Pattern, 6
Orlando, Florida, 62, 63, 71, 129, 148, 157, 173, 233, 240, 290, 298
Ospreys, 7
Pacific Ark, 27, 58, 67, 108, 109

INDEX

Phobos, ix, 143, 244
Pitholem, 86, 287
plasma consciousness, 35, 85
Pleiades, 169, 291
pomegranate, 212, 214, 224
Project Blue Beam, 32, 33
Quartermaster, 329, 331
Reptilians, 48, 99
Ringmakers, of Saturn, x, 322, 324, 325
Rise of the Red Dragon, 341
Rodrigues, Tony, 317, 318
Ruezo Zanrico, 293
Russia, 72, 98, 132, 134, 138, 168, 241, 248, 292, 293
Salla, Dr Michael, iv
sarcophagi, 97, 98
Sasquatch, 48, 228, 346
Saturn, viii, x, 2, 3, 119, 261, 272, 273, 274, 278, 281, 314, 316, 318, 319, 320, 321, 322, 323, 324, 325
Secret Space Program, 341
Seeder extraterrestrials, 74, 86, 282, 284, 285, 287, 291, 293, 301, 310, 346
sleeping giant, 2, 3, 127, 203, 216, 217, 218, 309, 311, 312
Solar Warden, 312, 334
South Africa, 13
space arks, i, iii, iv, 1, 3, 27, 33, 60, 70, 72, 73, 86, 93, 116, 132, 142, 164, 165, 291, 331
Space Command, 29, 146, 261, 267, 320
Space Force, US, 341
Spain, 52, 184
Special Operations Command, 328

Star Nations News, 319
Star Trek, 341
Sumerian, 47, 188, 217, 310
supersoldier, 245
Switzerland, 168
Tampa, 3, 85, 146, 148
Teleportation, 94, 95, 96, 97, 213, 304
Thor Han Eredyon, 74, 297, 301, 309, 310, 312, 320, 334
TR-3B, antigravity craft, 42, 81, 82, 83, 102, 182, 245, 254
Tree of Life, 127, 203, 216, 217, 218
UFOs, 250, 342
UNCLOS, 132, 133, 134, 135, 136, 139
Underwater cultural heritage, 136, 137
Unidentified Aerial Phenomena, UAP, 346
United States, iv
US Air Force Secret Space Program, 341
US Air Force, USAF, ix, 3, 20, 21, 51, 105, 147, 159, 196, 240, 245, 257, 290, 298, 301, 328, 329, 334
US Army, 1, 3, 5, 15, 19, 21, 27, 51, 62, 63, 75, 82, 100, 105, 112, 135, 145, 146, 178, 196, 204, 217, 233, 235, 239, 240, 256, 262, 281, 302, 307, 327, 328, 329, 330, 331, 333, 335, 339, 341, 346
US Navy, 16, 81, 135, 301
USAF secret space program, 51
Vimanas, 292
Vulcans, 155, 288
Washington, DC, 339
White Hats, v, 20, 328, 333

Endnotes

[1] Interview originally published in audio form on September 25, 2023, see Michael Salla, "JP Update 25 – Returning the Activation Jewel to the Atlantic Space Ark," https://exopolitics.org/jp-update-returning-the-activation-jewel-to-the-atlantic-space-ark/ (accessed 8/7/2024).

[2] See Michael Salla, *US Army Insider Missions 2: Underground Cities, Giants & Spaceports*, pp. 315-36.

[3] See Michael Salla, *US Army Insider Missions 2,* Figure 40, p. 323.

[4] See Michael Salla, *US Army Insider Missions 1: Spoce Arks, Underground Cities & ET Contact*, pp. 229-44.

[5] The incident involving JP's first visit to the Navy platform is described in Michael Salla, *US Army Insider Missions 1,* pp. 74-78.

[6] See Michael Salla, *US Army Insider Missions 1,* pp. 291-308.

[7] Jean Charles described his Bahamas experience in an interview titled, "Space Arks & Halls of Records in Antarctica, Giza, Tibet & Bahamas," Exopolitics Today April 29, 2022. https://michaelsalla.com/2022/04/29/space-arks-halls-of-records-in-antarctica-giza-tibet-bahamas/ (accessed 8/7/2024).

[8] See Michael Salla, *US Army Insider Missions 2*, p. 191.

[9] See Abraham (Avi) Loeb and Sean M. Kirkpatrick: "Physical Constraints on Unidentified Aerial Phenomena," draft version March 7, 2023, https://lweb.cfa.harvard.edu/~loeb/LK1.pdf (accessed 6/16/2023).

[10] Elena Danaan's description of the arrival of the Seeders is described in her book, *The Seeders: The Return* of the Gods (2022)

[11] JP describes the events before and during his mission to Ganymede in US Army Insider Missions 1, pp. 158-60.

[12] See Michael Salla, *US Army Insider Missions 2*, pp. 316-35.

[13] See Leonard Lewin, *Report from Iron Mountain* (1967).

[14] Interview originally published in audio form on October 23, 2023, see Michael Salla, "JP Update – Mission to Nordic Inner Earth Civilization for Life Extension Technology," https://exopolitics.org/jp-update-mission-to-nordic-inner-earth-civilization/(accessed 8/7/2024).

[15] See Sunbow Truebrother, *The Sasquatch Message to Humanity* (2020)

[16] See Michael Salla, *US Army Insider Missions 1*, pp. 48-50 & 74-78.

[17] See HeartMath Institute, "Exploring the Little Brain in the Heart: A Journey Into Heart-Brain Communication," https://www.heartmath.org/articles-of-the-heart/little-brain-in-the-heart/ (accessed 8/9/2024)

[18] See Michael Salla, *Rise of the Red Dragon: Origins & Threat of China's Secret Space Program* (Exopolitics Consultants, 2020).

[19] Interview originally published in audio form on December 7, 2023, see Michael Salla, "Nordics take control of Space Arks & Reveal Disturbing

Timelines to Earth Alliance –JP Update No.27," https://exopolitics.org/nordics-take-control-of-space-arks/ (accessed 8/9/2024).
[20] See Michael Salla, *US Army Insider Missions 1*, pp. 127-28.
[21] See Michael Salla, *US Army Insider Missions 2*, pp. 287-313.
[22] Elena Danaan, *Seeders* (2023) p. 314.
[23] See "Crashed UFO recovered by the US military 'distorted space and time,' leaving one investigator 'nauseous and disoriented' when he went in and discovered it was much larger inside than out, attorney for whistleblowers reveals," Daily Mail, https://www.dailymail.co.uk/news/article-12175195/Crashed-UFO-recovered-military-distorted-space-time.html (accessed 8/10/2024).
[24] Alex Collier initially raised these concerns in an online webinar and later in a private meeting on Zoom. See https://www.crowdcast.io/c/alex-collier-question-answer-webinar-45 (accessed 9/1/2024).
[25] The Atlantic Space Ark missions involving the Aztec Indians were discussed in *US Army Insider Missions 1*, pp. 229-68.
[26] Interview originally published in audio form on December 14, 2023, see Michael Salla, "Nordics & Inner Earthers take control of Space Arks – JP's 6th Mission to Atlantic Space Ark," https://exopolitics.org/nordics-inner-earthers-take-control-of-space-arks/ (accessed 8/10/2024).
[27] See Michael Salla, "Visit to Area 51 and Ancient Alien EM Fusion Containment Engine – David Adair Interviews Part 2," https://exopolitics.org/visit-to-area-51-and-ancient-alien-em-fusion-containment-engine/ (accessed 8/10/2024).
[28] The Atlantic Space Ark missions involving the Aztec Indians were discussed in *US Army Insider Missions 1*, pp. 229-68.
[29] The Narkina 5 image appeared in Episode 8 in the First Season of Andor. More info at: https://starwars.fandom.com/wiki/Narkina_5_Imperial_Prison_Complex (accessed 8/11/2024).
[30] See Morgan Dunn, "The Disturbing Story Of The Heart Attack Gun Invented By The CIA During The Cold War," https://allthatsinteresting.com/heart-attack-gun (accessed 8/11/2024).
[31] See Michael Salla, *US Army Insider Missions 1*, pp. 72-74.
[32] Interview originally published in audio form on January 11, 2024, see Michael Salla, "
"Strange Medical Experiments & More on Nordics Taking Over Space Arks – JP Update #28 https://exopolitics.org/strange-medical-experiments-more-on-nordics-taking-over-space-arks/ (accessed 8/16/2024).
[33] Stephen Hawking's holographic presentation in Hong Kong is available at: https://youtu.be/suS_flYdKEc?si=xKtujq43TewI7u91 (accessed 8/12/2024).
[34] I discussed this incident in Michael Salla, *US Army Insider Missions 2*, pp. 5-32.

ENDNOTES

[35] The photo and incident were discussed in *US Army Insider Missions 1,* 127-36.

[36] See David Jacobs, *The Threat: Revealing the Secret Alien Agenda* (Simon & Schuster, 2008) and Budd Hopkins, *Intruders: The Incredible Visitations at Copley Woods* (Random House, 1987)

[37] Elena Danaan reports being rescued by Nordic-looking extraterrestrials belonging to the Galactic Federation of Worlds. See Elena Danaan, *A Gift from the Stars* (2020).

[38] "United Nations Convention on the Law of the Sea," https://www.un.org/depts/los/convention_agreements/texts/unclos/unclos_e.pdf (accessed 8/16/2024).

[39] Encyclopedia Britannica, "Exclusive Economic Zone," https://www.britannica.com/topic/exclusive-economic-zone (accessed 8/14/2024).

[40] "United Nations Convention on the Law of the Sea," https://www.un.org/depts/los/convention_agreements/texts/unclos/unclos_e.pdf (accessed 8/16/2024).

[41] JP was first taken to the floating donut shaped platform in 2016 but was denied access. See *US Army Insider Missions 1*, pp. 63-65.

[42] "United Nations Convention on the Law of the Sea," https://www.un.org/depts/los/convention_agreements/texts/unclos/unclos_e.pdf (accessed 8/16/2024).

[43] "Convention on the Protection of the Underwater Cultural Heritage," https://www.unesco.org/en/legal-affairs/convention-protection-underwater-cultural-heritage?hub=412 (accessed 8/14/2024).

[44] "Convention on the Protection of the Underwater Cultural Heritage," https://www.unesco.org/en/legal-affairs/convention-protection-underwater-cultural-heritage?hub=412 (accessed 8/14/2024).

[45] "Convention on the Protection of the Underwater Cultural Heritage," https://www.unesco.org/en/legal-affairs/convention-protection-underwater-cultural-heritage?hub=412 (accessed 8/14/2024).

[46] See Jake Carter, "Puzzling Monoliths on Mars and Phobos: Who Put That There?" https://anomalien.com/puzzling-monoliths-on-mars-and-phobos-who-put-that-there/ (accessed 8/16/2024).

[47] See Jake Carter, "Puzzling Monoliths on Mars and Phobos: Who Put That There?" https://anomalien.com/puzzling-monoliths-on-mars-and-phobos-who-put-that-there/ (accessed 8/16/2024).

[48] Interview originally published in audio form on February 1, 2024, see Michael Salla, "Nordic Extraterrestrial Assimilation Program – JP Update #30" https://exopolitics.org/nordic-extraterrestrial-assimilation-program-jp-update-29/(accessed 8/16/2024).

[49] Dr. Jim Segala discussed the appearance of gamma wave EM energy in relation to alien abductions in the following News Nation interview, "Scientist explains unidentified aerial phenomena | Reality Check, "

https://www.youtube.com/watch?v=HoIaVvU-VUE&t=838s (accessed 8/16/2024).

[50] Elena Danaan's 2022 presentation titled: "Technologies of the Star Nations," is available online at: https://vimeo.com/ondemand/gsic2022/ (accessed 8/17/2024).

[51] The mission is described in US Army Insider Missions 2, pp. 261-86.

[52] Howard Menger, From Outer Space (Pyramid Books, 1959) pp. 66-67.

[53] See Stefano Breccia, 50 Years of Amicizia – 'Friendship' (2013)

[54] See Michael Salla, "Russian Prime Minister claims extraterrestrials live among us," https://exopolitics.org/russian-prime-minister-claims-extraterrestrials-live-among-us/ (accessed 8/17/2024).

[55] JP's YouTube account is https://www.youtube.com/@JPjpJP1 and his Instagram channel is: https://www.instagram.com/jp.missions/ /(accessed 8/28/2024).

[56] See Michael Salla, *Galactic Diplomacy: Getting to Yes with ET* (Exopolitics Consultants, 2013).

[57] Interview originally published in audio form on February 20, 2024, see Michael Salla, "The Ancient Underground Castle with Gold Plates – JP Update #31," https://exopolitics.org/the-ancient-underground-castle-with-gold-plates//(accessed 8/17/2024).

[58] See Michael Salla, *US Army Insider Missions 1*, 309-310.

[59] See "The Ancient Underground Castle with Gold Plates – JP Update #31" https://www.youtube.com/watch?v=FC24YtGpuig&t=2257s

[60] Interview originally published in audio form on April 1, 2024, see Michael Salla, "JP Update #32 – Ant People Move to New Realm after Ant King Transitions & Sleeping Giant Awakens," https://exopolitics.org/jp-update-31-ant-people-move-to-new-realm-after-ant-king-transitions-sleeping-giant-awakens/ (accessed 8/20/2024).

[61] See, "The Complete Guide to Land Navigation with the Military Grid Reference System," https://www.itstactical.com/skillcom/navigation/the-complete-guide-to-land-navigation-with-the-military-grid-reference-system/ (accessed 8/20/2024).

[62] The video, "JP Instructions for making Longevity Drink," is available at: https://rumble.com/v5cywka-jp-instructions-for-making-longevity-drink.html (accessed 8/20/2024).

[63] The video, "JP Instructions for making Longevity Drink," is available at: https://rumble.com/v5cywka-jp-instructions-for-making-longevity-drink.html (accessed 8/20/2024).

[64] See Michael Salla, *US Army Insider Missions 2: Underground Cities, Giants & Spaceports*, pp. 5-32.

[65] See Michael Salla, *US Army Insider Missions 2*, pp.51-71.

⁶⁶ See Heinrich Kusch and Ingrid Kusch, *Geheime Unterwelt: Auf den Spuren von Jahrtausende alten unterirdischen Völkern: Das Vermächtnis der Jahrtausende alten unterirdischen Völkern* (Authal Verlag, 2024)
⁶⁷ See "JP Instructions for making Longevity Drink," https://rumble.com/v5cywka-jp-instructions-for-making-longevity-drink.html (accessed 8/20/2024).
⁶⁸ See Radu Cinamar, *Inside the Earth: The Second Tunnel*, (Sky Books, 2019).
⁶⁹ JP's previous mission to an underground spaceport in the Alabama area was discussed in US Army Insider Missions 2: Underground Cities, Giants and Spaceports, pp. 209-36.
⁷⁰ Interview originally published in audio form on April 29, 2024, see Michael Salla, "JP Update #33 – Nordics Training International Military Pilots to fly Flying Saucers," https://exopolitics.org/jp-update-33-nordics-training-international-military-pilots-to-fly-flying-saucers/ (accessed 8/21/2024).
⁷¹ See Jennifer Young, "This Map Shows The Shortest Route To 6 Of Alabama's Most Incredible Caves," https://www.onlyinyourstate.com/trip-ideas/alabama/most-incredible-caves-map-al/ (accessed 8/21/2024).
⁷² See Michael Salla, *Antarctica's Hidden History: Corporate Foundations of Secret Space Programs* (Exopolitics Consultants, 2018).
⁷³ There are three popular caves only several miles from Huntsville Alabama that show the potential for large caverns to be found or created in the area. See "Huntsville's Historic Three Caves https://www.huntsville.org/blog/list/post/huntsvilles-historic-three-caves/ (accessed 8/21/2024).
⁷⁴ See Michael Salla, *US Army Insider Missions 2*, pp. 209-36.
⁷⁵ See Michael Salla, *US Army Insider Missions 1,* pp. 128-34.
⁷⁶ NBC News, "NASA panel studying UFO sightings says stigma and poor data pose challenges."
 https://news.yahoo.com/nasa-panel-studying-ufo-sightings-201913754.html (accessed 6/30/2023).
⁷⁷ See Michael Salla, *US Army Insider Missions 1, p*p. 40-41
⁷⁸ See Elena Danaan, *Area 51: Conversations with Insider Stephen Chua* (2023).
⁷⁹ See "The Artemis Accords, https://www.nasa.gov/artemis-accords/ (accessed 8/21/2024).
⁸⁰ I discussed the Jupiter Accords in *Galactic Federations, Councils and Secret Space Programs* (2022) pp. 137-58. See also, Elena Danaan, *We Will Never Let You Down* (2021) pp. 236-53.
⁸¹ See "International Lunar Research Station," https://www.unoosa.org/documents/pdf/copuos/2023/TPs/ILRS_presentation202 30529_.pdf (accessed 8/21/2024).
⁸² Transcribed from podcast clip found at: "Tucker Carlson shares bizarre tale of troops dying from UFO encounters," https://www.militarytimes.com/off-

duty/military-culture/2023/03/16/tucker-carlson-shares-bizarre-tale-of-troops-dying-from-ufo-encounters/ (accessed 6/30/2023).

[83] See Travis Walton, *Fire in the Sky* (De Capo Press,1997).

[84] For a description of Hyrdonium, see "The Hydronium Ion," https://tinyurl.com/2tfa2fk8 (accessed 8/22/2024).

[85] See Vanessa Romo and Bill Chappell, "U.S. recovered non-human 'biologics' from UFO crash sites, former intel official says," https://www.npr.org/2023/07/27/1190390376/ufo-hearing-non-human-biologics-uaps (accessed 8/22/2024).

[86] Interview originally published in audio form on July 18, 2024, see Michael Salla, "Testing for a Covert Space Mission to Saturn – JP Update #35," https://exopolitics.org/testing-for-a-covert-space-mission-to-saturn-jp-update-35/ (accessed 8/22/2024).

[87] See Michael Salla, *US Army Insider Missions 1*, p. 312.

[88] Private Whatsapp communication from Jean Charles Moyen received on August 22, 2024.

[89] See Michael Salla, *US Army Insider Missions 1*, pp. 158-60 & 171-94.

[90] See Michael Salla, *US Army Insider Missions 2*, pp. 91-116.

[91] See Faruk Imamovic, Former NOAA Head Claims Contact with Non-Human Intelligence," https://www.msn.com/en-us/news/world/former-noaa-head-claims-contact-with-non-human-intelligence/ar-AA1lUHr1 (accessed 8/23/2024).

[92] See https://exopolitics.org/jp-articles-photos-videos/ (accessed 8/23/2024).

[93] Also discussed in the original interview was JP's upcoming first public appearance from September 27-29 in Westminster, Colorado, at the Galactic Spiritual Informers Connection. More info is available at: https://www.galacticspiritualinformers.com/ (accessed 8/23/2024).

[94] Interview originally published in audio form on July 22, 2024, see Michael Salla, "Roundtable on Space Arks, Sleeping Giants, ET Assimilation and Mysteries of Saturn," https://exopolitics.org/roundtable-on-space-arks-sleeping-giants-et-assimilation-and-mysteries-of-saturn/ (accessed 8/22/2024).

[95] David Adair described his meeting with Pitholem and how it downloaded into him in the following interview, "Visit to Area 51 and Ancient Alien EM Fusion Containment Engine – David Adair Interviews - Part 2" https://exopolitics.org/visit-to-area-51-and-ancient-alien-em-fusion-containment-engine/ (accessed 8/22/2024).

[96] These first three Atlantic Space Ark missions are described in *US Army Insider Missions 1*.

[97] See Radu Cinamar, *Inside the Earth: The Second Tunnel - Transylvania Series Book 5* (Sky Books, 2019**)**.

[98] See Jimmy Joe, "Apkallu: The Legendary Sages Who Brought Divine Wisdom to the World," https://www.timelessmyths.com/mythology/apkallu/ (accessed 8/24/2024).

ENDNOTES

[99] See "Aruna (Hinduism)" https://en.wikipedia.org/wiki/Aruna_(Hinduism) (accessed 9/5/2024)

[100] See "Aruna (Hinduism)" https://en.wikipedia.org/wiki/Aruna_(Hinduism) (accessed 9/5/2024

[101] Norman Bergrun, *Ringmakers of Saturn* (The Pentland Press Ltd.,1986) p. viii

[102] Luis Elizondo, *Imminent: Inside the Pentagon's Hunt for UFOs* (Harper Collins, 2024). Ch 10, Time stamp 8:30.

[103] Private conversation with George Kavassilas on August 22, 2024.

Made in the USA
Middletown, DE
10 September 2024

60781443R00205